DOPPLER SPACE TIME

GERALD GRUSHOW

STARWAY SCIENTIFIC PRESS

VIRGINIA BEACH, VIRGINIA

DOPPLER SPACE TIME

By Gerald Grushow B.S.E.E. (Summa cum Laude)

Published by:
Starway Scientific Press
P.O. Box 62174
Virginia Beach, Virginia 23466
Email: starway@infi.net

All rights reserved. No part of this book may be reproduced or transmitted in any form or by any means, electronic or mechanical, including photocopying, recording or by any information storage and retrieval system, without written permission from the author except for the inclusion of brief quotations in a review.

Copyright © 1999 and 2000 by Gerald Grushow

Printed in United States of America

Library of Congress Control Number: 00-090688

Book is available directly from publisher.

Publisher's Cataloging-in-Publication
(Provided by Quality Books, Inc.)

Grushow, Gerald.
    Doppler space time / Gerald Grushow. -- 1st ed.
    p. cm.
    Includes index.
    ISBN: 0-9703301-1-1

    1. Electromagnetic theory. 2. Unified field theories. 3. Space and time. 4. Relativity (Physics) I. Title.

QC670.G78 2000   530.14'1
                              QBI00-720

ACKNOWLEDGEMENTS

The Author gratefully acknowledges the love and patience of his wife Karen, who has sacrificed greatly during the writing of this book over the last twenty years. Without her help and support this great effort would not have been possible.

The Author gratefully acknowledges the help and encouragement of his children, Debbie, Jeffery, and Melissa. In addition, the author thanks his Nephew Martin for his helpful suggestions over the years.

A special thanks goes to Andy Cusumano, a mathematician friend and expert in converging and diverging series, and progressions who proof-read the original manuscripts and who again proof read this final version.

A special thanks goes to Melvin Bridges, a family friend and graphic design artist for his outstanding cover design.

To Mike

Hope you enjoy this book

[signature]

Dec. 2, 2000

ABOUT THE AUTHOR

The Author was born Dec. 24, 1938 in Brooklyn, N.Y. He went to P.S. 25 in Bedford Styvesant. He later went on to Brooklyn Technical High School. The Author took the Electrical Course at Brooklyn Tech and got the only 100% grade in the six hour Electrical Comprehensive exam in 1956.

Upon graduating High School, he went to work for Con Edison of New York in 1956 as a Technician while going to Polytechnic Institute of Brooklyn at night. He was given the Engineering job of planning for the installation of 4000 volt substation distribution systems for the Bronx N.Y. He also designed the 13,000-volt distribution system for Freedom Land Amusement Park, which later became Coop City.

In 1961 he went to work for the Dept. of Water Supply, Gas, and Electricity at the Manhattan Municipal building as an Associate Engineer. He redesigned the lighting and power distribution for the Hutchinson River Parkway, the Long Island Expressway, and the 59th Bridge.

In 1966 he graduated college after 10 years of night school with a B.S.E.E. (Summa cum Laude). In 1967 he went to work for Sperry Corporation of Great Neck N.Y. There he worked on various military systems including the WDE system.

In 1970, he went to work for the Port of N.Y. Authority. He designed lighting Towers for Kennedy and LaGuardia Airports. He also designed control systems for the fuel pumping station at Kennedy.

In 1972 the Author returned to Sperry to work on the Mk92 Dutch radar system. Later he joined a Gun fire control system group and designed a control system for the 5-inch Navy guns. In later years, the system evolved to be the SDC/GMP used on the Belnap and other Aegis ships.

Over the years he worked on the Nexrad system, the Trident sonar system, the Polaris missile system, the ring laser gyro, and the infrared detection system for the Radar Research and Development Department. The author retired in 1993.

The seven years of retirement has enabled the Author to complete his work on the Dot theory, gravity, and Doppler space-time. It enabled him to present the Dual Doppler Earth as well. The Author is now 62 years old. This book is the Author's last great effort and adventure. It took 20 years to reach this point. Now the Author leaves to others the knowledge gained from this adventure into the unknown but not unknowable.

DOPPLER SPACE-TIME

PREFACE

In 1981, the Author embarked upon a study of the Physics of the Universe in order to find the various missing links between physics and electrical theory. The initial study consisted of numerical analysis of the constants of the universe using a hand calculator. Rough concepts and ideas concerning the basic structure of matter and light accompanied the numerical study for several years.

From 1981 to 1983 there was some initial breakthroughs. The Author determined that the entire universe was electrical in nature and that it was composed of a multiplicity of only two things. The Author called these dots. In 1986, an initial manuscript was produced and sent to the U.S. Copyright office. By 1988 additional chapters were completed and copyrighted. Some of the basic equations remain till today however much is new or completely revised.

The Dot theory in Chapter 1 of the book was worked on since its conception in 1981. It was modified several times but remains as a Method of solving the Universe. It is a way of looking at the universe from a purely electrical point of view.

Einsteinian space-time always suffered the necessity of Doppler corrections for Radar systems. The clock paradox and the lack of any reference system also illustrated the necessity for a true space-time analysis. In addition, the failure to find an explanation for the double slit experiment led physicists to conclude that no such explanation was possible. These problems necessitated a new general space-time, which would not be subject to such problems.

The initial space-time analysis started within the Dot theory. The Author in 1999 decided to add Doppler Radar principles to the Dot theory to encompass complete space-time. The principles of GG/MM Doppler space-time were then formulated.

The book contains the physics of the universe from the point of view of the fundamental particle or dot. It also contains a dual mathematical approach based upon the principles of Doppler space-time. The reader can now look at the universe from the simple Dot theory, or from the more complex Doppler space-time theory.

Gravity is described by the Dot theory. Gravity is also described alternatively in the Doppler space-time theory. The Dot theory is a simple way of looking at the universe. It enables the reader to grasp the entire universe in his mind from the smallest sub-particle to the radius of the universe. It provides a working model of the universe. Doppler space-time enables the reader to understand the mechanism of gravity, inertia, and the dot itself. Both theories enable the reader to see an entire solution to the universe. As the book progresses, the dot becomes the Doppler dot. Initially it is merely a focal point. Later it is understood as a very interesting Doppler universe in itself. Each dot is its own universe.

The Dot theory and the Doppler space-time theory will enable the production of new and better sources of energy and rocket thrusters. A non-radioactive atomic generator will be built someday using ordinary simple fuels. There will be no radioactive waste. A proton thruster will also be built someday. This will convert the proton into pure photon energy. This will be the power source for future space ships and enable them to move toward the speed of light. The outcome of the understanding of the Dot theory will be the production of cheap/safe power and energy for future man.

The most important breakthrough is the conversion from mass to charge momentum in which the principles of a unified field theory are formulated. This is shown in Chapter 2.

The first thirteen chapters are devoted to the physics of the Universe. In Chapter 13, the Dual Doppler Earth is described in equations. The meaning of the Dual Doppler Earth requires a philosophical explanation, which is undertaken in chapters fourteen through sixteen.

The Author presents a scientific philosophical viewpoint in the middle ground between the thoughts of believers and those of atheists. This enables readers of various faiths, or no faith at all to view the Doppler Earth from many different viewpoints. The most important thing is the understanding of the physics involved in the life/death process.

In Chapters 17-19, a preparation for a trip to Alpha Centauri is planned and then attempted. This is one method by which man can survive a disaster upon the Earth.

The reader will come to understand the problems of near light speed space travel and the lessons of Doppler space-time will become self evident during the voyage.

PHYSICS CONTENTS

ONE	**THE DOT THEORY**	**1**
1-1	POSTULATES OF THE DOT THEORY	1
1-2	THE MODEL UNIVERSE	2
1-3	GRAVITATIONAL BACKPRESSURE	5
1-4	THE DOT STRUCTURE OF THE UNIVERSE	7
1-5	THE MASS OF A DOT	8
1-6	THE MASS OF THE UNIVERSE	11
1-7	THE DOT DC VOLTAGE CALCULATION	13
TWO	**CONVERSION OF MASS TO CHARGE**	**17**
2-1	CONVERSION THEORY	17
2-2	GENERAL MASS TO CHARGE CONVERSIONS	19
2-3	THE LIGHTSPEED EQUATIONS	24
2-4	THE GRAVITATIONAL CONSTANT	25
THREE	**THE OSCILLATING UNIVERSE**	**29**
3-1	THE INDUCTIVE/CAPACITIVE OSCILLATOR	29
3-2	VARIABLE SPACE-TIME MATHEMATICS	30
3-3	MAGNETIC & ELECTRIC FIELDS OF THE UNIVERSE	33
3-4	MINIMUM RADIUS OF THE UNIVERSE	35
3-5	THE CLOCKS OF THE UNIVERSE	36
3-6	THE LIGHT SPEED HYSTERESIS LOOP	38
FOUR	**PROTONS, NEUTRONS, ELECTRONS, & PHOTONS**	**41**
4-1	THE P* SUBPARTICLE	41
4-2	PROTON GENERATION	42
4-3	THE ELECTRON IN NEUTRON'S ORBIT	45
4-4	ELECTRON DE BROGLIE WAVELENGTH	48
4-5	THE BOHR ORBIT	49
4-6	THE FINE CONSTANT CALCULATION	51
FIVE	**AC MODELS OF PARTICLES & PHOTONS**	**53**
5-1	PROTON & NEUTRON AC OSCILLATIONS	53
5-2	PHOTON'S AC OSCILLATION	56
5-3	ELECTRON'S AC OSCILLATION	60
5-4	ELECTRON FORCES	65
5-5	PROTON/PROTON BINDING	66
SIX	**GG/MM/DOPPLER LINEAR SPACE TIME**	**71**
6-1	GG/MM/DOPPLER SPACE-TIME PRINCIPLES	71
6-2	LINEAR SPACE TIME	75
6-3	DOPPLER MASS CENTER OF GRAVITY	83
6-4	DOPPLER RELATIVITY REFERENCE SYSTEM	85
6-5	LENGTH & TIME INVARIANCE	88
6-6	CHARGE INVARIANCE	89

PHYSICS CONTENTS

SEVEN	GG/MM/DOPPER ORBITAL SPACE TIME	91
7-1	GG/MM/DOPPLER ORBITAL MASS EQUATIONS	92
7-2	INERTIA IN M/M SPACE-TIME	96
7-3	THE DOPPLER LENGTH	100
7-4	CLASSICAL ORBITAL SPACE-TIME MASS INCREASE	101
EIGHT	ROTATIONAL SPACE TIME	103
8-1	ROTATION IN M/M SPACE TIME	103
8-2	THE GYROSCOPE IN SPACE TIME	108
8-3	ORBITAL CLOCKS	110
NINE	SPACE & SPACE TIME	115
9-1	SPACE ETHER	115
9-2	DOUBLE SLIT EXPLANATION	118
9-3	DOUBLE SLIT IN SPACE TIME	120
9-4	THE ADAPTABLE PHOTON	122
9-5	SPACE TIME QUANTUM MECHANICS	123
9-6	BOHR ORBIT IN SPACE TIME	124
9-7	ELECTRIC CHARGE IN SPACE TIME	127
9-8	FASTER THAN LIGHT	130
TEN	THE FM DOPPLER MODEL UNIVERSE	133
10-1	FM DOPPLER SPACE TIME	133
10-2	MASS OF UNIVERSE AT BIG BANG	136
10-3	THE FM DOPPLER UNIVERSE	138
10-4	PHASE TIME	141
ELEVEN	THE LAWS OF GRAVITY	145
11-1	THREE DIMENSIONAL FORCES IN SPACE TIME	145
11-2	FORCES ACTING ON BODIES	147
11-3	THE GRAVITATIONAL FORCE	152
11-4	GRAVITY IN SPACE TIME	157
TWELVE	UNIVERSAL GRAVITATION	163
12-1	THE LAWS OF UNIVERSAL GRAVITATION	163
12-2	GRAVITATIONAL RULES	170
12-3	DOPPLER GRAVITY	177
THIRTEEN	THE DUAL DOPPLER EARTH	179
13-1	THE DOPPLER IMAGE	179

PHYSICS & PHILOSOPHY CONTENTS

FOURTEEN	**THE ORIGIN OF THE UNIVERSE**	**183**
14-1	IN THE BEGINNING	183
14-2	THE OSCILLATING UNIVERSE	184
14-3	FROM BIG BANG TO LITTLE BANG	187
14-4	THE PERFECT STATE OF THE UNIVERSE	188
14-5	THE MULTI-LIGHT-SPEED SOLUTION	190
FIFTEEN	**THE BIRTH OF THE DOPPLER LIFE-FORCE**	**191**
15-1	THE CYCLES OF THE UNIVERSE	191
15-2	THE PRODUCTION OF THE DOPPLER MIND	192
15-3	THE EVOLUTION OF THE LIFE-FORCE	194
15-4	THE SPACE TIME BRAIN	196
15-5	THE GALAXY LIFE-FORCE	197
15-6	MULTI-LIGHT-SPEED CONSIDERATIONS	200
SIXTEEN	**SPIRITUAL PHYSICS**	**201**
16-1	ORBITAL SPIRITUAL PHYSICS	201
16-2	ROTATIONAL SPIRITUAL PHYSICS	207
16-3	PRODUCTION OF ONE CELL LIFE	209
16-4	THE STEPS OF EVOLUTION	211
16-5	THE FINAL PHYSICAL MAN AND WOMAN	213
16-6	MAN AND WOMAN IN A SPIRITUAL PARADISE	214

PHYSICS & SPACE TRAVEL CONTENTS

SEVENTEEN	**PREPARATION FOR TRIP TO ALPHA CENTAURI**	**217**
17-1	THE PROTON THRUSTER ENGINE	217
17-2	THE NON-RADIOACTIVE ATOMIC GENERATOR	219
17-3	THE ABSOLUTE VELOCITY SENSOR	220
17-4	SPACESHIP REQUIREMENTS	222
EIGHTEEN	**TRIPS AROUND THE SOLAR SYSTEM**	**225**
18-1	INITIAL TRIAL RUNS AND TESTS	225
18-2	TRIP TO THE MOON	226
18-3	TRIPS TO MARS AND VENUS	228
18-4	TRIP TO PLUTO	228
18-5	TRIP TO REACH MAXIMUM SPEED	228
18-6	PREPARATION FOR THE VOYAGE	230
NINETEEN	**THE VOYAGE TO THE NEW EARTH**	**233**
19-1	THE WAR ON THE EARTH	233
19-2	THE TRIP TOWARD ALPHA CENTAURI	234
19-3	RETURNING TO THE NEW EARTH	236

INTRODUCTION

This book starts in Chapter 1 with the equations of a simple sphere of light. This provides us with a set of normalized electrical equations, which permit a model of the universe to be produced. In this model, the universe is one gigantic electromagnetic field. The field has a huge number of focal points called dots, which have a DC charge of $+Q_D$ or $-Q_D$.

There are only two different things in the entire universe, the plus dot and the minus dot. The characteristics of the dots are such that large quantities of the dots form the charges Q and minus Q. The electron is not a cloud of just minus dots. It is a sandwich of plus and minus dots with an excess of minus dots equal to the charge Q. The same is true for the proton. The proton has an enormous amount of plus and minus dots with an excess of positive dots, which produce the positive charge Q. The proton has approximately 1836 times as many dots as the electron and the difference of mass relates to the total number of dots in each particle.

The electrical world of magnetism and static charges is caused by the electrical characteristics of the dots. The effects of space and time upon the dots cause the mechanical world. Repetitive patterns or motions of the dots cause steady state conditions, which permit space-time memory effects. The image of the present and the image of the past of a split milli-second ago tend to form current loops, which react with each other.

When you move an object you struggle against the image of the object of a split milli-second ago. The present image of an atom produces a counter-force when the atom is moved from its past image of a split milli-second ago. Ampere's laws of force apply. Mass and inertia are electrical properties, which appear to us as mechanical properties.

In Chapters 1-13, the Dot theory, the conversion from mass to charge momentum, gravity, and Doppler space-time will be discussed. In Chapters 14-16, the philosophical implications of Doppler space-time will be investigated. Finally in Chapters 17-19, a space adventure toward Alpha Centauri will be planned and carried out. In the end, the spacecraft will return to the New Earth. Then higher man will emerge from the children of the space men.

Let us now begin the exciting adventure with the Dot theory.

CHAPTER 1- THE DOT THEORY

SECTION 1-0 INTRODUCTION

In this chapter, the Dot theory will be introduced and a model of the universe will be presented. This will enable us in future chapters to understand gravity, space and time, the binding energy of protons and atoms, and the oscillation of the universe itself. We begin with the postulates of the Dot theory.

SECTION 1-1 POSTULATES OF THE DOT THEORY

The Dot theory states that the entire universe is a huge singular electromagnetic field. It states that everything in the universe is electrical in nature, and that mass is an electrical property of space and time. The Dot theory states that everything in the universe is composed of a multitude of only two things, a plus dot and a minus dot. The electron, the proton, the neutron, and the photon are all composed of dots. The same is true of all the antiparticles. The only difference between things is the particular geometric configuration of each particle or photon.

The Dot theory states that the electron is composed of plus and minus dots with an excess of minus dots equal to the total charge Q= 1.602E-19 coulombs. The proton contains approximately 1836 times as many plus and minus dots as the electron and has an excess of positive dots equal to the charge Q= 1.602E-19 coulombs.

The dot is a quanta of the electromagnetic field. It is an electro-photon or charged photon. The dot can be considered a focal point of the electromagnetic field. This is the point of a current flow between dual halves of the universe, which are separated in space-time by a tiny distance.

The basis of the Dot theory is that the universe is a singular electromagnetic field. The dot charge, and or the dot current will drop as the universe expands. Likewise, when the universe contracts, the dot charge, and or the dot current increase. In the process, the universe oscillates from a small space-time radius to a very large space-time radius as will be calculated in chapter 3.

The first effort of the book is to quantize the dots. A method has been chosen based upon a simple model universe, which has interesting Thevenin equivalent characteristics. The equivalents are used to approximate an Einsteinian space-time universe. In general, for the Einsteinian universe, every point is the center of the universe. For the Dot theory model universe, a corollary to the Einsteinian solution is that the universe looks like a perfect sphere to every point in the universe. This permits easy calculations of the dot characteristics since the dot sees a very simple universe of radius R_U.

In the Dot theory, the model is used to produce electrical equations for the rest mass/energy of the electro-photon dot. The dot is a Doppler effect dot. It reaches to the radius of the universe moving at the speed of light there and it also exists within the atom as a hole between the universes with a current flow moving at the speed of light between the universes. The dot or the current flow always moves at the speed of light C but it can form standing waves of energy within matter. The dot then becomes part of the standing wave.

The dot contains a current flow between halves of the universe at the dot point. Plus dot current flows one way and minus dot current flows the other way. This discharges both the positive universe and the negative universe simultaneously. The universe then looks like two batteries, which are short-circuited.

The dot energy will be quantized and the exact number of dots per neutron will be calculated. The radius of the universe and the time of universe is readily found by the Dot theory and matched to the astronomical data using numeric analysis of the constants of the universe. The numeric analysis seeks simple equations, which match the universe. The numeric theory states that the universe is tied together algebraically by simple expressions and simple numbers such as 4, π, e, 137, etc.

In general standard electrical and mechanical equations and data are used throughout the book. The Dot theory provides the quantization of the energy of the universe. The Dot theory equations provide the basis for the conversion from Mass to Charge in chapter 2.

The work from Chapter 3 on uses standard electrical and mechanical theory. Let us now begin with the model universe.

SECTION 1-2: THE MODEL UNIVERSE

Let us start out with some very simple algebraic equations as we search for an understanding of the relationships between the electrical world and the mechanical world of the universe. In order to understand how the universe functions, it is necessary to start out with a simple model of the universe and then explore the model and expand upon it until it matches the actual measurements obtained in the laboratory. The universe is modeled upon huge numbers of expanding light spheres.

Every point in the universe is the center of the universe. All individual electromagnetic spheres originating at a dot form a total complex spherical shape such that all spheres intersect all other spheres in the universe. The spheres all interlock in three dimensions of distance and three dimensions of time such that the universe always looks like an absolutely perfect sphere at every point in the universe.

Let us look at a single sphere of light expanding from a pinpoint as representing the electromagnetic universe. As the light expands, the radius of the universe also expands and a standard ruler and a standard time clock expand as well. Time clocks slow down and rulers stretch as the universe expands. Notice that if the universe expands at light speed, it expands much faster when it is small then when it is large. The rate of expansion or the acceleration of the light sphere is:

$$A_U = (C^2)/R_U \qquad (1\text{-}1)$$

Equation 1-1 states that an expanding sphere of electromagnetic light energy has acceleration, A_U of meters/second^2 that is simply the speed of light, C squared divided by the radius, R_U of the light sphere. An idealized universe at every point follows the form of equation (1-1)

This simple equation forms the basis of an idealized universe comprised of a single huge electromagnetic field, which experiences time and distance elongation.

The total force acting upon the model universe is simply the mass of the Universe M_U times the acceleration of the universe A_U:

$$F_U = M_U A_U = (M_U C^2)/ R_U \qquad (1\text{-}2)$$

The expansion of the universe is driven by the difference between the electromagnetic and electrostatic fields. When the universe expands the electromagnetic field is slightly greater than the electrostatic field. From an AC viewpoint, the magnetic field during expansion leads the electrostatic field by a slight angle.

When the universe contracts the electrostatic field is slightly greater than the electromagnetic field. From an AC viewpoint, the magnetic field lags the electrostatic field during contraction. Therefore an electrical hysteresis loop exists within space and time which causes all the light spheres to expand and contract.

As the universe expands, there is a very slight drop in light speed as will be calculated in chapter 3. We have a light speed hysteresis loop as will be explained in that chapter.

In this book we look at things from both an electrical and a mechanical viewpoint. The universe can be described either from electrical equations or mechanical equations.

From a mechanical viewpoint, the differential driving force acting upon the idealized expanding universe is the loss of mass per unit time. From an electrical viewpoint, the differential driving force is the loss of charge per unit time.

From a mechanical viewpoint:

$$F_U = -C\, d(M_U)/ d(T_U) \qquad (1\text{-}3)$$

In equation 1-3 we see that the driving force acting upon an expanding electromagnetic universe is the light speed C times the derivative of the mass of the universe with respect to the time of the universe. Notice that the expanding bubble of the electromagnetic field has less and less force acting upon it as it expands more and more toward infinity since the mass is decreasing toward zero. Correspondingly, the charge is decreasing to zero as well.

The radius of the universe R_U is related to the time of the Universe by the following:

$$R_U = C\, T_U \qquad (1\text{-}4)$$

The radius of the universe is simply the speed of light times the time of the universe since the pinpoint at big bang. The pinpoint is not a singularity. From the outside it may appear as a single point. From the inside it is the same as the number of dots in the universe, thus it is a multiplicity.

The energy of the model universe is simply the mass of the universe times the speed of light squared. It is also the force operating upon the universe times the radius of the universe. Thus:

$$E_U = F_U R_U = M_U C^2 \qquad (1\text{-}5)$$

Notice that all these equations are very simple general equations, which describe an expanding light sphere. Yet, they will produce accurate results when the concepts are applied to the standard equations of physics for all the detailed calculations as will be seen in the sections to follow.

If we set equation (1-3) equal to equation (1-2) and solve using equation (1-4) we get:

$$M_U C^2/R_U = -C\, d(M_U)/d(T_U) = -C^2\, d(M_U)/d(R_U) \tag{1-6}$$

In addition since the net or differential driving function is the loss of mass as the universe expands:

$$M_U R_U = \text{Constant} \quad \text{(Kilogram Meters)} \tag{1-7}$$

One solution to Equations 1-6 and 1-7 is an exponential function. In later chapters and sections an exponential sinusoidal solution will also be presented.

The simple solution to an expanding light sphere is

$$M_U = M_O\, e^{-x} \tag{1-8}$$

$$T_U = T_O\, e^{x} \tag{1-9}$$

$$R_U = R_O\, e^{x} \tag{1-10}$$

$$C_U = C_O \tag{1-11}$$

where X is a driving function which varies from minus infinity to plus infinity.

In equation (1-8) we see that the mass of the universe decreases as the driving function (x) of the universe increases. In equation (1-9) we see that the time of the universe expands as the driving function (x) of the universe increases. Finally in equation (1-10) we see that the distance of the universe expands as the driving function of the universe expands. Thus as the light sphere increases, both the ruler and the clock expand for a constant light speed universe as per equation (1-11). The driving function can be a simple linear function. It could also be a sinusoidal function. Thus our space and time could be driven by an inner sinusoidal function as will be explained in Chapter 3.

Since both charge and mass decrease as the universe expands, the following equations also apply:

$$Q = Q_O\, e^{-x} \tag{1-12}$$

In equation 1-12 we see that charge decreases as an ordinary exponential. It gets smaller as both time and distance increase. This will be fully explained in chapter 2.

$$U_O = U_{Oo}\, e^{2x} \tag{1-13}$$

In equation 1-13, the electrical permeability U_O increases to the second power as both time and distance increase.

$$K = K_O\, e^{2x} \tag{1-14}$$

In equation 1-14, the Coulomb's constant K increases to the second power as both time and distance increase. Likewise for the electrical permitivity:

$$\varepsilon_O = 1/K = \varepsilon_{Oo}\, e^{-2x} \tag{1-15}$$

In equation 1-15 we see that the electrical permitivity constant ε_0 decreases as the universe expands. The speed of light is:

$$C = 1 / (U_0 \varepsilon_0)^{1/2} \qquad (1\text{-}16)$$

In equation 1-16 we see that the speed of light remains constant as the universe expands since the electrical permeability constant increases while the electrical permitivity constant decreases. The small light-speed hysteresis loop will be explained in chapter 3. The impedance of the universe is:

$$Z_U = Z_{U_0} e^{2x} \qquad (1\text{-}17)$$

In equation 1-17 we see that as the universe goes toward infinity, the impedance of the universe also goes toward infinity and it becomes an open electrical circuit. As the universe heads toward a pinpoint at minus infinity, the impedance of the universe goes to zero. Thus it becomes an electrical short circuit.

From this we see that the universe is an inductive/capacitive oscillator which varies from a short circuit at big bang to an open circuit at infinity. The small light-speed hysteresis loop prevents compression to zero and expansion to infinity for the exponential sinusoidal solution. The cycle time of the universe and the differential light speed will be calculated in Chapter 3.

From these simple relationships of an expanding light sphere, everything can be understood. It is necessary to pinpoint on the model specific measurements, which will lock in the exact values. This will be shown in the chapters to follow.

SECTION 1-3: GRAVITATIONAL BACKPRESSURE

In this section we will take a very brief look at the gravitational forces operating upon the expanding electromagnetic sphere of light.

The mass of the universe decreases with the distance the universe occupies as postulated in section 1-1. As the universe expands a backpressure due to the loss of mass with increased distance occurs. This is also due to the loss of charge as distance expands. This backpressure will tend to squeeze matter together. This will be discussed in detail in Chapter 2 and later chapters. Basically the backpressure can be described as Ampere's law of current loops. As the universe expands galaxies, stars, and planets will form due to the backpressure. Conversely, during the contraction of the Universe, gravity is negative and all planets and stars are destroyed. This will be explained later.

Let us now formulate some equations to yield the mass of the model universe. For an ordinary balance of mechanical forces, the force of attraction between two bodies would be:

$$G M M / R^2 = M V^2 / R \qquad (1\text{-}18)$$

In equation 1-18 we see that the gravitational force is equal to the centrifugal force in an ordinary planetary motion equation. When we consider the entire universe to be an expanding interlocking set of a huge number of light spheres of electromagnetic energy, a Thevenin equivalent of the universe can be produced. A simple force equivalent of the set of light spheres would be the force of two universes separated by a distance equal to the radius of the universe.

We can now equate the force of expansion with the gravitational force of the universe. The gravitational force will be the backpressure force. If you consider the universe to be composed of waves, then the backpressure on the waves will force the material world together. Thus:

$$F_U = M_U (C^2)/R_U = G M_U M_U / R_U^2 \qquad (1\text{-}19)$$

In equation (1-19) we see that the force driving the expansion of the electromagnetic field depends upon the mass, M_U of the universe and the speed of light, C squared divided by the radius, R_U of the universe.

On the other side of the equation we have a simple Thevenin equivalent of the contracting forces acting upon the universe. Solving for G:

$$G = (C^2) R_U / M_U \qquad (1\text{-}20)$$

In equation 1-20 we see that the electrical gravitational backpressure field constant G equals the speed of light squared times the radius, R_U of the universe and divided by the mass, M_U of the universe.

Since the light speed is basically constant, and since the radius of the universe expands with time and the mass of the universe decreases with distance, the gravitational constant G increases as the square of the radius of the universe. Since the mass decreases, the term GMM is a constant except for non-linearities or a slight hysteresis loop, which will cause the universe to oscillate.

If we look at equation 1-19 however, we see that the gravitational force decreases as the square of the distance with time since the mass of the universe decreases while the radius of the universe increases. Thus at full expansion, the gravitational forces weaken as the square of the radius of the universe. Therefore, stars and planets will explode. The protons will do no better since as will be shown in Chapter 5, they experience a slowing of their internal AC oscillations. Thus protons and electrons experience their own red shift effect, as do photons. They will be destroyed as we head toward maximum expansion. This is called the little bang where the entire universe returns to electro-photon energy and only homogeneous patterns of the dots remain.

In general, the atomic binding forces, which depend upon the DC Charge Q, weaken as time goes by. This causes non- radioactive atoms to become more and more radioactive as time goes by. Right after the big bang many atoms existed which do not exist today. As time goes by more and more atoms become radioactive. Eventually lead will become radioactive and in the future even oxygen will become radioactive. In the end of the process, the protons and electrons are also destroyed. Thus at full expansion everything is destroyed. At full expansion, we have only a homogeneous electromagnetic field left and this field will then collapse.

As the field collapses, the electrostatic attraction becomes greater that the electromagnetic repulsion. This insures the complete collapse of everything we know. Thus on the reverse cycle of the universe, no stable matter exists and it is a pure electro-photon universe which collapses toward a multiple pinpoint.

The universe will not reach a perfect multiple pinpoint size. It will reach a minimum size, which will be calculated in a chapter 3. Then it will explode again and a new universe will be born.

In this section the formula for the mass of the idealized universe was computed using a Thevenin equivalent of the universe. Various other methods will be used to produce idealized formulas to calculate the number of dots in the neutron and in the universe, etc. These methods serve to produce a physically realizable universe from which the protons, electrons, neutrons, photons, gravity, and space-time can be understood.

SECTION 1-4: THE DOT STRUCTURE OF THE UNIVERSE

In this brief section, let us understand the very simple nature of the universe. The entire universe is composed of only two different things. The first is a plus dot and the second is a minus dot. According to later calculations in this chapter, there are 1.46193E125 dots per universe. These calculations are based upon a normalized universe and everything is accordingly normalized. As more data is analyzed, corrections to the figures are possible in the future. However the workings of the universe do not change if corrections are applied. The changes are the dots per neutron, the radius of the universe, and the mass of the universe.

There are two different dots, which could be considered the focal points of the electromagnetic field of the universe. Each dot has a plus charge, Q_D or a minus charge, $-Q_D$. Clouds of dots make up the charge Q=1.602E-19 coulombs. Each dot has a local DC charge of 1.42186E-60 coulombs, which will be calculated later in this chapter. The dot also has a charge of 1.602E-19 coulombs at the radius of the universe. Therefore it is zero size and almost infinite size simultaneously. This is due to the Doppler effect of a charge moving at the speed of light C.

The local charge Q is made up of large numbers of both plus and minus dots with an excess amount of either positive or negative dots depending upon whether the object is a proton or an electron. The same is true for the antiparticles as well. In addition to the local charge Q_D of the dot, there is a local dot current flow I_D. There are mechanical properties of moving groups of dots and space itself, which cause mass, inertia, and gravity. This will be discussed later. In effect each dot has a local energy E_D, an equivalent rest mass M_D, a charge Q_D, a current I_D, a capacitance C_D, and an inductance L_D.

The charge Q_D with its equivalent rest mass M_D and current I_D comprise the entire structure of the universe. It is a purely electrical universe and everything is made from the dots. The dots can be considered multidimensional Doppler space-time entities.

The dots represent actual current flow between positive and negative halves of the universe. They are perfect point current sources. On each side of the universe, they appear to come from nowhere and go nowhere. The plus dot current can be said to flow toward us and the minus dot current can be said to flow away from us. However, it is all one sandwich of a positive/negative universe.

In general, the dots exist basically uniformly everywhere and are felt everywhere up to the radius of the universe. They are little bits and pieces of the discharging electromagnetic field itself. The universe is a dual electromagnetic field and the dots are the lowest active quanta of charge/energy of the field itself.

At the level of the dots, only rest mass exists. The dots are thus electro-photons or DC current flows. They have no mass in themselves but repetitive patterns of dots in synchronous motion produce steady state forces, which give rise to mass and inertia. The ordinary photons moving at the speed of light do not possess a repetitive image and thus appear mass-less. The proton and electron have repetitive patterns within themselves and produce synchronous images of their present and their past. They have both mass and inertia. The dot is an electro-photon and is free to form mass or to form photons. The only real difference is that the photons do not produce a repetitive relatively stationary image, and therefore do not obey Ampere's law of forces for current loops, which requires repetitive images.

The two lead balls in the laboratory are pushed together because of the expansion of the universe. This produces the universe of today and the universe of yesterday which are a few microseconds apart. The expanding electromagnetic field of each electron, proton, and neutron, plus all the hydrogen Bohr orbits and higher more complex atomic Bohr orbits produce current loops which create a magnetic force directed toward the center of all atoms and particles.

This backpressure which is Ampere's law of current loops is the cause of universal gravitational. The expanding universe causes positive gravity whereas the contracting universe causes negative gravity. All this will be explained in the sections and chapters to follow.

Photons travel at light speed when free but drop speed slightly when in a strong gravitational field, which increases the electric permitivity and magnetic permeability constants. The speed of light is slightly higher between galaxies than within a galaxy or near a star. This causes the light to bend around the star, as the photon becomes part photon part mass as will be explained in chapter 9 section 9-2. In addition, the red shift will be fully explained in chapter 5 section 5-2.

Two solutions for the variation of charge and mass with time are presented in this book. One is an e^x solution and one is an $e^{x \sin x}$ solution. The most likely solution is the exponential sinusoid because of the hysteresis loop as will be shown later. However both solutions are very similar because the maximum to minimum ratio of the universe is very large.

SECTION 1-5: THE MASS OF A DOT

In this section, we will look at the relationships between the electrical world and the mechanical world in a very simple manner. The charge Q will be used and explored as the source of the conversion. There will be a direct connection between the dot charge Q_D and the dot rest mass M_D.

In Section 1-2, it was postulated that the mass of the universe decreases as the radius of the universe increases. This is a premise of the Dot theory. The universe is composed of a large number of dots of local rest mass M_D and local charge Q_D. The dot also has the charge Q at the radius of the universe. The mass and charge of each dot is also decreasing as the universe expands. There is a DC current flow out of the charge Q at R_U similar to that of a parallel plate capacitor where the plates are expanding. In addition to the strong electrical coulomb DC forces acting in the universe, there are also strong magnetic forces acting in the universe.

These strong forces are basically equal to each other. During the expansion of the universe the magnetic forces are slightly stronger whereas during the contraction of the universe, the electrostatic forces are stronger. In addition, there is a net hysteresis loop, which produces the gravitational field. This is the differential electric/magnetic force field.

Let us look at the DC electrical field of a standard expanded charge Q at the surface of the model universe which is a perfect sphere moving at the speed of light. We find locally that:

$$V_D = K Q/R_U \tag{1-21}$$

In equation (1-21) we see that the voltage of a dot here caused by a charge Q located at the radius of the universe is very small but not zero. We have a finite universe of radius R_U, and there is a little voltage left over at the dot, which is the center of the universe. The dot looks like the charge +Q or –Q at the radius R_U. Likewise as we look at each positive charge Q of the proton, there is a little positive voltage left over at the radius of the universe. For the electron, a minus voltage also reaches the radius of the universe.

The universe is balanced with the charge Q at R_U and at our particles as well. We then have an electric battery for the universe where the current flows through the dots.

The outer sphere of the universe at radius R_U is a perfect conductor. There is a short circuit at the distance R_U everywhere. It is a virtual zero or ground plane of the universe. The same is true at every local point in the universe. Every point in the universe sees a ground plane a distance R_U away and also in the zone between the positive dot flow and the negative dot flow. This ground plane is halfway in the hysteresis loop in the distance dimensions of the universe.

Let us now look at the dot current flow. This dot current flow is not an ordinary current flow such as with an electron. It is the slow discharge of space-time similar to the decrease of charge of a parallel plate capacitor when the plates are slowly moved apart. In effect, the dot current is the discharge of space itself. The dot current flow is the flow across the barrier between the positive voltage universe and the negative voltage universe as the universe expands. The universe is a space-time capacitor, which charges and then discharges.

We can relate the dot current to the dot voltage by the impedance of space, which is the intrinsic impedance of electrical theory. The impedance of the universe Z_U is

$$Z_U = 4 \pi K / C \tag{1-22}$$

$$Z_U = 4 \pi K T_U / R_U \tag{1-23}$$

where K is the Coulomb's constant, T_U is the time of the universe, and R_U is the radius of the universe.

In order for the voltage at the neutral conducting sphere at the radius of the universe to be zero, there must be a dot current flow equal to the dot voltage divided by the impedance of the universe.

$$I_D = Q / (4 \pi T_U) \tag{1-24}$$

The charge Q could be thought of as the peak of a very tall mountain that exists at the radius of the universe. Over a very long period of time the distance to the mountain increases toward infinity and the peak of the mountain disappears. This is caused by a current flow. The current flow is similar to ordinary electrical capacitive circuits in which the plates are separated, and the charge on both plates' decrease. It is $V_D \, [d(C_D)/d(t)]$ where the change of capacitance (C_D) is related to the time of the universe.

The current flow can also be expressed as:

$$I_D = QC/(4\pi R_U) \tag{1-25}$$

In Equation 1-25 we have the charge Q at the distance R_U, moving away at the speed of light and causing the spherical current I_D. The current flow at the distance R_U is identical with the current flow at the local dot itself. The dot power flow into the local short circuit is the dot voltage times the dot current. Thus:

$$P_D = V_D I_D = K Q Q/(4\pi R_U T_U) \tag{1-26}$$

Likewise the dot power is the square of the dot current times the impedance of the universe.

$$P_D = (I_D)^2 Z_U \tag{1-27}$$

$$P_D = K Q Q C/(4\pi R_U^2) \tag{1-28}$$

The dot energy is the dot power times the time of the universe:

$$E_D = K Q Q/(4\pi R_U) \tag{1-29}$$

Finally the dot mass is the dot energy divided by the speed of light squared.

$$M_D = K Q Q/(4\pi R_U C^2) \tag{1-30}$$

Equation (1-30) gives us the basic conversion equation from mass to charge in a very simplified way. Later more detailed calculations will pinpoint the exact conversion. However this method permits us to look at the mass of a dot M_D and compare it to the coulomb constant K, the charge Q, the radius (R_U) of the universe, and the speed of light, C. It is an equation which matches the mechanical world to the electrical world at the units of kilograms. The mass is basically caused by the dot current flow. It is moving charge momentum as will be seen in Chapter 2.

The force acting on a dot is the energy of the dot divided by the radius of the universe. Thus:

$$F_D = E_D/R_U \tag{1-31}$$

$$F_D = KQQ/(4\pi R_U^2) \tag{1-32}$$

We can now relate the mass of the dot to the charge of a dot by means of the magnetic permeability constant U_O of free space. Using (ε_O) for the electrical permitivity constant we get:

$$\varepsilon_O = 1/(4\pi K) \tag{1-33}$$

$$C = 1/(\varepsilon_O U_O)^{1/2} \tag{1-34}$$

$$K = (U_O C^2)/4\pi \tag{1-35}$$

$$M_D = U_O Q Q/(4\pi R_U 4\pi) \tag{1-36}$$

We see that the Mass of a dot is related to the permeability U_O, the charge Q to the second power and the inverse of the radius of the universe. From equation 1-36 it is clear that mass is a magnetic effect.

SECTION 1-6: THE MASS OF THE UNIVERSE

The mass of the universe for the model universe can be calculated using Equation 1-20. Hence

$$M_U = (R_U C^2)/G \tag{1-37}$$

where M_U is the mass of the universe in Kg for the MKS system of units, R_U is the radius of the universe in meters, and G is the gravitational constant.

This equation is for a light sphere expanding from a pinpoint at the speed of light. It is a solution to an e^x curve. It is also applicable for an $e^{\sin(x)}$ solution

We can solve for the mass of the model universe if we know the radius of the universe or the time of the universe from the pinpoint. The same is true for a universe, which oscillates from e^{-x} a minimum to e^{x} a maximum since the time would be T_O and normalized to the point where $e^x = e^0 = 1$. We can normalize our solution to the time equal to zero. In this case, everything before us is negative time leading to minus infinity while everything ahead of us is positive time leading to plus infinity. In this way we can study a universe which comes from minus infinity and goes to plus infinity. It is the exponential function e^x that permits easy normalization for the model of the universe.

We need to normalize the time of the universe to some quantity. We assume that we are at Time zero on the e^x curve. The astronomers look at the Hubble line for their cosmic time scale. The Hubble line can be represented as the loss of energy of a photon per unit distance and not the Doppler red shift. The far stars are moving away from us as the universe expands but this is a common mode occurrence as the ruler expands as well. The expansion of the ruler negates the Doppler red shift but there is also a loss of oscillation energy of the photon as the dot charges within the photon decrease and the photon AC internal oscillation slowly dies out.

Alternately, the internal oscillation could be said to negate the slowing time clock and then the red shift would be the Doppler effect. It is just a different way of looking at it. For an expanding universe with expanding time clock, a second reason is always necessary for the red shift. Thus both the Doppler and the slowing of the AC oscillation are necessary to offset the ruler expansion.

A premise of the Dot theory is that coulombs decrease as the time increases. Thus the quantity coulomb second is a constant. This shows itself up in the slowing of all clocks in the universe and the slowing of the internal photon oscillation for the red shift. It also shows up in the slowing of the internal electron oscillation and the slowing of the internal proton oscillation as will be explained in Chapter 5. Thus:

$$\text{Coulomb Second} = \text{Constant} \tag{1-38}$$

We can then normalize the time of the universe as follows:

$$4 \pi Q T_U = 1^* = 1 \text{ (coulomb sec)} \tag{1-39}$$

In equation 1-39, Q is the standard charge of 1.60218E-19 coulombs and T_U is the time of the universe and 1* represents unity in coulomb seconds. Using equation 1-39 we normalized the time of the universe to be 4.96682E17 or 15.75 billion light years. The actual time can always be modified slightly. However when you deal with a normalized e^x function you can always describe things in terms of a 15.75 billion light year universe. This time was chosen to approximately agree with the Hubble astronomical data. It also agrees with simple numerical analysis of the various constants of the universe. In a later chapter a more exact method of finding T_U will be presented. However, the results are very close to each other. Therefore to a reasonable degree of Engineering accuracy (one percent), equation 1-39 is acceptable for the model.

The time of the universe then becomes:

$$T_U = T_O e^{(t/T_O)} = 4.96682E17 \text{ seconds} \tag{1-40}$$

In equation 1-40 we see that if today t=0, the universe is T_O or 15.75 billion years old, and for a pure e^x solution this is always true. Corrections can always be made because the time (t) could be a million years or a billion years since the time when T_U equaled T_O. However this doesn't change how the universe works. It merely changes the exact point on the curve we are on. We may not have the exact measurements but we know the general answer anyway.

In the exponential sinusoidal solution the same is true, we can correct the initial time and set the exact phase angle we are at. Then the entire solution becomes known. This would require great scientific analysis of all the astronomical data. It is a job for the future.

The radius of the universe is simply:

$$R_U = C T_U \tag{1-41}$$

This calculates to be:

$$R_U = 1.48901E26 \text{ Meters} \tag{1-42}$$

Using equation 1-37 and G=6.67260E-11, the normalized mass of the universe is:

$$M_U = 2.00559E53 \text{ kilograms.} \tag{1-43}$$

We can now calculate the equivalent mass of a dot from equation 1-30.

$$M_D = K Q Q / 4\pi R_U C C \tag{1-44}$$

In equation 1-44 we repeat the terms instead of using the square for clarity. Thus the Mass of a dot M_D is equal to the coulomb constant K times the charge Q squared divided by the speed of light C squared, divided by the radius of the Universe R_U, and finally divided by 4π.

Using K=8.98756E9, Q=1.60218E-19, C=2.99792E8, R_U=1.48901E26, and π=3.14159, we get:

$$M_D = 1.37188\text{E-}72 \text{ Kg} \qquad (1\text{-}45)$$

The number of dots in the universe is the mass of the universe divided by the mass of each dot:

$$N_U = M_U / M_D \qquad (1\text{-}46)$$

$$N_U = 1.46193\text{E}125 \qquad (1\text{-}47)$$

In equation 1-47 we have broken the universe down into a large amount of dots of electromagnetic energy quanta by the normalization process. At this point in time the entire model universe has been quantized into tiny bits and pieces of energy and charge.

The number of dots in the neutron is the mass of the neutron, 1.67493E-27kg divided by the mass of a dot:

$$N_N = M_N / M_D \qquad (1\text{-}48)$$

$$N_N = 1.22090\text{E}45 \qquad (1\text{-}49)$$

There are a huge number of dots in one neutron. The number of neutrons in the universe can be calculated using the mass of the universe and dividing by the mass of the neutron. This is an equivalent number since a proton and electron plus some photon energy is considered a neutron.

$$\#N = M_U / M_N \qquad (1\text{-}50)$$

$$\#N = 1.19742\text{E}80 \text{ neutrons per universe} \qquad (1\text{-}51)$$

These are normalized numbers for the model. The engineer may be happy with these numbers while the scientist will want to know the exact answer. The engineer can build things without knowing all the exact details while the mathematician and scientist want the fine points of all the answers The object of the Dot theory is to provide a working model of the universe. In general most chapters in this book are independent of the Dot theory and rely upon standard scientific data and theory. Thus the Dot theory is only one part of the book.

SECTION 1-7 THE DOT DC VOLTAGE CALCULATION

The dots of our existence can be considered as focal points of the electromagnetic field with current flow. The dot always travels at the speed of light at the distance R_U. Locally they move or form standing wave patterns. The dot always has momentum of $M_D C$. It always has an energy of $M_D C^2$. The dot is also part of an AC oscillation as well, such as the internal electron oscillation, the proton oscillation, and or the photon oscillation. This will be calculated in Chapter 5.

The local charge of a dot has not been calculated yet. Equation 1-36 gives us its normalized mass but we do not know its charge. The dot charge at R_U is Q, but the local charge is a Doppler effect, which is discussed in Chapter 9. For the moment let us calculate the dot charge by a simple method.

It takes a certain number of dots to make up the charge Q within the proton or electron. There are many charges (Q's) within the proton. There are at least 1835 plus and minus standard electrical charges within the proton. Thus if you chop the proton up you can get at lease 917 negative charges and 918 positive charges of standard charge Q equal to 1.6022E-19 coulombs. The standard charge Q is a geometric structure of many dots whose total plus and minus charges exceeds that of the measured electron charge.

In effect, the proton can be chopped up into positive electrons and negative electrons. However, once you produce the proton's geometric structure all you have is a complex combination of plus and minus dots. These dots will make everything but they will rotate and spin differently for different geometric configurations. Thus matter and anti-matter are all composed of the same dots. However, the geometry of the dots and their collective oscillations, and spins produce the various primary masses.

The dot equations are as follows:

$$\text{Voltage dot} = KQ/R_U \tag{1-52}$$

Equation 1-52 is a repeat of equation 1-21 from section 1-5. The dot voltage is equal to the electric constant K times the charge Q divided by the radius of the universe R_U. That means the universe acts like a perfect sphere of radius R_U with a charge Q upon the surface of that sphere, when we look at the universe from a mirror image viewpoint.

The Dot theory states that the universe is a perfect sphere for every single dot in the universe. It could be viewed as a perfect sphere with us near the center, or as an Einsteinian perfect sphere. It doesn't make any difference since if we appear as the center of an Einsteinian space-time universe, the universe itself always looks like a perfect sphere to us. Thus the equations will always be the same.

Let us now calculate the proton voltage. The radius of the proton is identical with its deBroglie wavelength as will be shown in Chapter 4. The voltage of the proton can be calculated readily. Since K=8.98756E9, Q=1.60218E-19, and R(proton) = 1.32142E-15, we can calculate the voltage as follows.

$$\text{Voltage proton} = 1{,}089{,}710 \text{ volts} \tag{1-53}$$

The proton voltage is such that once you enter the proton radius you are charged to a little over one million volts. The dot voltage from a charge Q at a distance of R_U equal to 1.48901E26 is:

$$V_{Dot} = KQ/R_U = 9.67065E{-}36 \tag{1-54}$$

The proton voltage is equal to the sum of the voltage due to the plus dots, minus the sum of the voltage due to the minus dots.

$$\text{Voltage (proton)} = (\text{Number of Excess Plus dots}) \cdot V_D \tag{1-55}$$

From this we can calculate the number of excess dots which make up the single charge Q, and the dot charge Q_D.

$$\text{Number excess plus dots} = V_{proton} / V_{Dot} \tag{1-56}$$

$$\text{\# dots for charge Q} = 1.12682\text{E}41 \text{ dots.} \tag{1-57}$$

$$Q_D = [Q / \text{\# dots per Q}] = 1.42186\text{E-}60 \tag{1-58}$$

The number of dots in the neutron is:

$$\text{Dots neutron} = 1.22090\text{E}45 \tag{1-59}$$

The corresponding number of dots for a particle is:

$$\text{Dots particle} = \text{Dots Neutron} \times \text{Mass Particle} / \text{Mass Neutron} \tag{1-60}$$

$$\text{Dots proton} = 1.21922\text{E}45 \tag{1-61}$$

$$\text{Dots electron} = 6.64008\text{E}41 \tag{1-62}$$

where $M_P = 1.67262\text{E-}27$, $M_E = 9.10939\text{E-}31$, $M_N = 1.67493\text{E-}27$, and $N_N = 1.22090\text{E}45$.

The number of charges of plus and minus Q within the proton is:

$$\text{Number of charges} = \text{Dots Proton} / \text{Dots per charge Q} \tag{1-63}$$

$$\text{Number of charges per proton} = 1.08200\text{E}4 \tag{1-64}$$

There are a little over ten thousand individual standard charges per proton. This assumes that the proton voltage remains perfectly flat after the radius R(proton), R_P. It may very well be that we have to reach into the root mean square radius of the proton, which is 0.7071 R_P to get the exact proton voltage. However for the model universe, the number of dots per unit charge in the above calculations is satisfactory.

The number of charges per electron is equal to the number of dots per electron divided by the number of dots for a charge Q; this assumes that the electron radius approximately equals the proton radius. Thus:

$$\text{Number of charges per electron} = 5.8928 \tag{1-65}$$

Equation 1-65 tells us that within the electron are the equivalent of 3 ordinary electrons and 2 positive electrons, and less than one electron's worth of photon energy in a configuration to produce gravitational mass. This does not mean that there are five separate particles within the electron. It only means that if you broke it apart you could produce five charged particles of value Q = 1.60218E-19 coulombs each. This of course would be quite unstable.

In general it is very difficult to break apart the electron. It is light. It will move fast. The proton is more stationary and thus can be broken apart more readily. Yet, it breaks into more massive parts that just the electron. However in the future man should be able to extract the proton's energy as pure photon energy.

The Dot theory does not go into the world of sub-particles. It is primarily concerned with the ultimate sub-particle, the dot that makes everything.

In this chapter the dot mass/energy has been quantized by equation 1-36 at the value of 1.37188E-72Kg. Most of atomic theory quantizes photon energy using Plank's constant h; the corresponding equivalent formula would be:

$$M_D = h / (8 \pi^2 (137.036) C R_U) \tag{1-66}$$

$$M_D = 1.37188\text{E-}72 \text{ Kg} \tag{1-67}$$

In Chapter 9, the Doppler charge will be discussed. For the moment, the ratio of the charge Q to the dot charge is:

$$Q/Q_D = R_U / R_P = 1.12683\text{E}41 \tag{1-68}$$

In equation 1-68 we see that the ratio of the charge Q and the dot charge Q_D is identical to the ratio of the radius of the universe and the radius of the proton. The reason for this is that if you move a charge at the speed of light it will look like a small charge to the left and a much larger charge to the right. This will be explained in chapter 9 once the Doppler space-time theory has been presented.

The dot current and other calculated values will be found in Appendix 1, Table of calculated values.

CHAPTER 2 – CONVERSION OF MASS TO CHARGE

SECTION 2-0: INTRODUCTION

In this chapter the simple equations of the Dot theory will be used to find the conversion from mass to charge. The universe is an electrical universe and the units of kilograms can be described electrically. At present the universe is specified in terms of kilograms, coulombs, meters, and seconds. There are many different names for things such as Amperes but Amperes are really coulombs per second. In the usual mechanical/electrical physics books, four units are required. Once we convert mass in terms of coulombs, meters, and seconds then only three units are required.

The mass to charge conversion is limited to the constraints of the universe that we live in. If the universe is expanding as per the Dot theory with increasing ruler and time clock while at basically constant light speed, then we have fewer conversion choices. In Section 2-1 Dot Theory conversion equations are used. In Section 2-2 some general mass to charge conversions are looked at. They represent a whole class of conversions but only those with a basically constant light speed solution are considered possible.

SECTION 2-1 CONVERSION THEORY

Most of the equations in Chapter 1, are a set of normalized equations for a model universe that expands at a basically constant light speed and with an expanding ruler and an expanding time clock. Some of the equations such as equation 1-34 are standard electrical equations, whereas others are for the normalized model universe. These equations serve to reduce the equations of physics into three variables of kilograms, meters, and seconds; or coulombs, meters, and seconds. Using these equations, it will no longer be necessary to use combined electrical and mechanical units.

In order to eliminate either kilograms or coulombs, it is necessary to produce a table of conversion. We can replace kilograms with some function of coulombs, meters, and seconds. Thus we can define mass in terms of charge and some power of light speed. Charts can then be produced using many combinations that match the equations in Chapter 1 as far as units are concerned.

The process for the production of the charts is readily accomplished. The comparison study of the charts takes many years to accomplish. Each chart will produce equations for the universe that may or may not represent reality. Fortunately the charts tend to be dual or sister solutions so things learned from an incorrect chart still applies to our universe. Thus all the solutions can be considered Sister solutions. However, only one solution is the actual solution and this is called the Sister 1 solution.

As a Sister 1 solution we could say that mass has the units of coulomb meters/second and that energy has the units of coulomb meters^3/seconds^3. This solution meets the Dot theory criteria of basically constant light speed since as meters increase and seconds increase, the light speed remains constant. The mass then only depends upon the coulombs, which decrease as the universe expands. This solution says that mass is a property of moving charge. It also says that energy is charge moving in a volumetric fashion.

For a Sister 2 solution, a dual solution could be looked at. Mass would be coulomb seconds/ meter and energy would be coulomb meters/ second. In this solution charge would be a property of a moving mass.

In general for a Dot theory solution, mass must equal charge times any power of light speed. Therefore,

$$M_D = Q_D C^n \qquad (2\text{-}1)$$

where C^n could be any positive or negative power of the light speed. The corresponding dot energy is:

$$E_D = Q_D C^{n+2} \qquad (2\text{-}2)$$

In equation 2-2 we see that once the form of the dot mass is chosen, the energy is C^2 higher since $E=MC^2$.

We can now produce a chart of the Sister 1 solution and the Sister 2 solution to see the various relationships. In general the Sister 1 solution states that:

$$M = Q C \qquad (2\text{-}3)$$

Equation 2-3 states that the unit of mass is charge times the speed of light. This basically states that a moving charge causes the property of mass. The dual solution or Sister 2 solution is:

$$Q = MC \qquad (2\text{-}4)$$

Equation 2-4 states that the unit of charge is mass multiplied by the speed of light. This basically states that a moving mass causes the property of charge.

The Author has investigated many other solutions over the years but the simplicity of equations 2-3 and 2-4 causes either one to be considered a most likely solution. However, the Sister 1 equations will be shown to match the physical universe more readily.

Charts of other solutions are easily prepared, However, once higher powers or square roots or cubes of the light speed occurs, they lack the simplicity of the Sister 1 solution or the dual Sister 2 solution.

The important thing in the method shown is that the equations presented so far are unit's equations in which only the three ingredients of kilograms, meters, and seconds, or coulombs, meters and seconds are used. Since the mechanical world interlocks with the electrical world at energy and force equations, it was never necessary to have so many different units. At most we only needed kilograms, coulombs, meters, and seconds. By specifying that the universe is completely electrical, only three units are needed.

Let us now look at a table of the two dual most likely solutions for the relationship between mass and charge.

The following table of units is made for the most likely Sister 1 solution with Mass = Charge times Velocity and the dual Sister 2 solution with Charge = Mass times Velocity.

TABLE 2-1: CONVERSION OF MASS TO CHARGE (MCS SYSTEM)

Quantity	Sister 1	Sister 2
System	MCS	MCS
Mass (M)	Cou Met/Sec	Cou Sec/Met
Charge (Q)	Coulombs	Coulombs(Cou)
Energy (E)	Cou Met^3/ Sec^3	Cou Met/Sec
Coulomb Const (K)	Met^4/ Cou Sec^3	Met^2/Cou Sec
Force (F)	Cou Met^2 / Sec^3	Cou /Sec
Momentum (MV)	Cou Met^2 / Sec^2	Coulombs
Plank's Const (h)	Cou Met^3 / Sec^2	Cou Met
Permeability (U$_O$)	Met^2 / Cou Sec	Sec/Cou
Permitivity (ε_o)	Cou Sec^3 / Met^4	Cou Sec/Met^2
Voltage (V)	Met^3 / Sec^3	Met/Sec
Current (I)	Cou / Sec	Cou/sec
Impedance (Z)	Met^3 / Cou Sec^2	Met/Cou
Grav Const (G)	Met^2 / Cou Sec	Met^4/CouSec^3
Power (P)	Cou Met^3 / Sec^4	Cou Met/Sec^2
Flux Density (B)	Met / Sec^2	1/met
Inductance (L)	Met^3 / Cou Sec	Met Sec/ Cou
Charge/Mass (Q/M)	Sec/ Met	Met/Sec
Capacitance (C)	Cou Sec^3/ Met^3	Cou Sec/Met

In the table the various quantities have been shown in the Meters Coulombs Seconds (MCS) system for the most likely Sister 1 and Sister 2 solutions.

SECTION 2-2: GENERAL MASS TO CHARGE CONVERSIONS

In Section 2-1, the Dot theory method of matching mass to charge was formulated. The basis of the method was that charge and mass both decreased with time or distance as the universe expanded. The relationship between mass and charge then became ratios of the speed of light to various positive and negative powers.

A more complete solution of various possibilities would be that mass also could vary with charge or current and a power of the speed of light. Thus more possibilities exist beyond the Dot theory although they become less probable.

Let us look at the simplest general possibilities. The following are possible:

Mass = Charge	(M=Q)	(2-05)
Mass = Current = Charge per second	(M=I)	(2-06)
Mass = Charge per meter	(MC=I)	(2-07)
Mass = Current x Velocity	(M=IC)	(2-08)
Mass = Charge /Velocity	(MC=Q)	(2-09)
Mass = Charge x Velocity	(M=QC)	(2-10)

We can now make charts of all these conversions by using the following standard equations:

$$\text{Force} = KQQ/R^2 \qquad (2\text{-}11)$$
$$h = \text{Energy} \times \text{Time} \qquad (2\text{-}12)$$
$$GMM = KQQ \qquad (2\text{-}13)$$
$$V = KQ/R \qquad (2\text{-}14)$$
$$U_0 \varepsilon_0 = 1/C^2 \qquad (2\text{-}15)$$
$$B = U_0 I/R \qquad (2\text{-}16)$$

Equations 2-11 through 2-16 are standard physics equations and enable us to produce charts of the various relationships in terms of meters, coulombs, and seconds for the MCS system. We can also chart the modified GG/MKS system of meters, kilograms, and seconds.

The solution for mass = charge, (Equation 2-05), tells us that a moving mass produces currents, and positive and negative magnetic fields from positive and negative dots. Since mass is loaded with dots, one characteristic of mass is identical to charge itself. However, when we look inside the proton, we see dots in constant motion. We see current flows and they look like current gyroscopes. Thus a primary characteristic of mass is not charge but it is related to the motion of charge. Although (mass = charge) falls within the Dot theory since both mass and charge decrease with an increasing universe, it is not a real possibility.

The solution mass = current, (Equation 2-06) is a characteristic of mass. If we look inside the proton, we do find currents within it. Yet, what give them the gyroscopic nature are not currents themselves but currents flowing in a circular path. This solution is not a Dot theory possibility since mass and charge do not track each other as the universe expands.

The next possibility for the primary electrical characteristic of mass is that momentum is current, (Equation 2-07). Moving masses do have moving dots and currents exist. That mass is a coulomb per meter can be considered a characteristic of mass. However, we are looking for a gyroscopic effect for the electrical equivalent of mass. This solution is not a Dot theory solution but it is interesting to study and compare to the other solutions.

The next possibility is that mass is current times velocity, (Equation 2-08). In this case, as the universe expands the mass drops and the charge remain constant. This solution has all independent units. No basic relationship between the gravitational constant and the magnetic field is self-evident. It does provides an alternate solution, however is not part of the Dot theory.

The next possibility is that mass equals charge over velocity, (Equation 2-09), or that a moving mass creates charge as a dual Sister 2 solution. It provides a good electrical to mechanical analogy with mass and capacitance being equal. It remains the secondary dual solution for comparison study and analogy since the differential equations of capacitive circuits are of identical form as those equations for mass.

Let us now look at the last solution, (Equation 2-10) which is the Sister 1 solution. In this solution a moving charge produces mass. Thus:

$$\text{Charge} \times \text{Velocity} = \text{Mass} \tag{2-17}$$

Equation 2-17 is for charge momentum. It is the electrical dual of a physical momentum. Thus:

$$\text{Mass Momentum} = \text{Mass} \times \text{Velocity} \tag{2-18}$$

$$\text{Charge Momentum} = \text{Charge} \times \text{Velocity} \tag{2-19}$$

We see in equations 2-18 & 19 that both mass and charge momentum's are similar quantities. Breaking down equation 2-19 into parts, we get:

$$[\text{Charge/second}] \times \text{Meters} = I \times R = \text{Mass} \tag{2-20}$$

In equation 2-20 we see that we have a current (I) operating at a radius R that gives us a current torque or charge momentum. This is a strong possibility for mass since we can see within the proton, currents flowing in circular paths and oscillating from inner radius to outer radius as well. This will be explained in chapter 5. We also see a root mean square current at a particular radius. Thus the current gyroscopic action within the proton adds credibility to equation 2-20 as being a primary conversion equation. Let us now produce a chart of the various relationships for this Sister 1 solution for both the electrical (MCS system) and the mechanical (GG-MKS system). In the later, the MKS system has been modified to show charge (Q) in terms of mechanical units where charge is kilogram seconds per meter.

TABLE 2-2: MASS TO CHARGE CONVERSION FOR SISTER 1 SOLUTION (M=QC)

Quantity	MCS System	GG-MKS System	Equalities
Mass (M)	cou met/sec	kg	
Charge (Q)	coulombs	kg sec/ met	
Charge/Mass (Q/M)	sec/met	sec/met	
Velocity	met/sec	met/sec	
Acceleration	met/sec^2	met/sec^2	**
Energy (E)	cou met^3/sec^3	kg met^2/sec^2	
Force (F)	cou met^2/sec^3	kg met/ sec^2	
Momentum (MV)	cou met^2/sec^2	kg met/sec	
Plank's Const. (h)	cou met^3/sec^2	kg met^2/sec	
Coulomb Const. (K)	met^4/cou sec^3	met^5/kg sec^4	
Permitivity ε_O	cou sec^3/met^4	kg sec^4 / met5	
Permeability (U$_O$)	met^2/cou sec	met^3/ kg sec^2	*
Grav. Const (G)	met^2/cou sec	met^3/kg sec^2	*
Voltage (V$_D$)	met^3/sec^3	met^3 / sec^3	
Current (I)	cou/sec	kg/met	
Impedance (Z)	met^3/cou sec^2	met^4/ kg sec^3	
Inductance (L)	met^3/cou sec	met^4 / kg sec^2	
Capacitance (C)	cou sec^3/ met^3	kg sec^4/ met^4	
Uo I	met^2/ sec^2	met^2/sec^2	
Flux Density (B)	met/sec^2	met/ sec^2	**

In Table 2-2 the gravitational constant has the same units as electrical permeability. This shows that gravity is a magnetic force. In addition flux density (B) has the same units as acceleration. This shows that an accelerating space-time electromagnetic field produces magnetic flux. Since voltage is cubic velocity, this shows that the moving universe generates voltage. Thus the motion of the electromagnetic field produces currents and voltages and magnetic flux as it expands or contracts.

The relationship between the gravitational constant and the electrical permeability using a best numeric fit is:

$$G = 16 \pi e U_O / (137.036)^3 = 6.6720E-11 \tag{2-21}$$

Equation 2-21 shows the exact relationship between the gravitational constant and the electrical permeability constant. It shows that the gravitational field is purely magnetic and that the conversion chart in Table 2-2 provides us with a very important conversion from mass to charge.

In the Sister 1 (MCS) solution mass per charge as per Equation 2-17, is a constant as the universe stretches out with both time and distance expanding. Energy is volumetric in both time and distance and drops as coulombs drop. Thus:

$$MV^2 = Q V^3 \tag{2-22}$$

In Equation 2-22, energy has the units of charge times velocity cubed. Likewise it is charge within the confines of cubic meters over cubic seconds. Thus energy is a volumetric space-time entity. It is charge oscillating in three dimensional distance (meters) volume over three dimensional time (seconds) volume.

$$MV = Q V^2 \tag{2-23}$$

In Equation 2-23, for momentum, coulomb meters squared per seconds squared would be charge oscillating in two dimensional distance (meters) area over two dimensional time (second's) areas. Thus all the various units mean something with respect to dimensions of velocity which will be explained in section 2-3.

The Sister 2 solution has the simplest units. However, the more complex units of the Sister 1 solution provide us with a greater understanding of space and time. Since there are three dimensions of light speed as will be explained later, there are three dimensions of distance and three dimensions of time also.

The fact that energy has the cubic units for the Sister 1 solution makes that solution more important. The Sister 2 solution is a great duality and since $M_N C = \pi Q$ approximately, it appeared most important originally. The Sister 1 solution matches at the gravitational constant / electrical permeability constant. However, that is what was sought since this is where the mechanical world matches the electrical world. Of course we must find what mass matches what charge. The product of QC is:

$$QC = 4.80321E-11 \tag{2-24}$$

In equation 2-24 QC is a large number. It appears to be a root mean square Doppler mass as will be discussed in Chapter 9. It is approximately the geometric mean between the mass of a dot and the mass of the universe.

The sister 1 solution enables us to see that mass is the little current gyroscope within the proton and electron. In the photon, you may get circular action perpendicular to the plane of motion but no circular action in the front to back region. Thus the photon will spiral and move forward and have no mass in the front to back direction. The photon will merely be a planar current loop which possesses no three dimensional gyroscopic ability. This will be explained in Chapter 7 of the book.

The five solutions investigated all are part of the general conversion equation from mass to charge. Thus:

$$GMM/R^2 = U_0 QV QV/ R^2 \qquad (2\text{-}25)$$

Equation 2-25 is a general unit's equation of standard physics that relates the gravitational force to the electrical magnetic attraction. The units are standard physics. The five most probable solutions to this equation are:

$G= U_0$ $\qquad$ $(M = QC)$ $\qquad$ (2-26)

$G= U_0 R$ $\qquad$ $(M = QC/R)$ $\qquad$ (2-27)

$G= U_0 C^2$ $\qquad$ $(M = Q)$ $\qquad$ (2-28)

$G= U_0 C^4$ $\qquad$ $(MC = Q)$ $\qquad$ (2-29)

$G = U_0 R^2 C^2$ $\qquad$ $(MC = QC/R)$ $\qquad$ (2-30)

If G and U_0 are reasonably true constants, they should not be a function of the distance R with respect to each other. Thus equations 2-27 and 2-30 are less likely to have much meaning. In addition, the distance R that would match G and U_0 is not of the order of the Bohr orbit or the proton radius. When we look at equation 2-29, we see that G would be of the same form as the electrical permitivity. In addition, if both M and Q were identical, we would be at a loss to explain all the properties of the universe.

The net result is that the Sister 1 solution of Equation 2-26 and the dual Sister 2 of equation 2-29 form the possible primary characteristics of the mass to charge conversion. Since the numeric relationship between G and U_0 in equation 2-21 is merely the $16 \pi e$ term and the 137.036 term, it appears as the most important solution. Of course this does not deny the importance of the dual solution of equation 2-29 since the masses are inter-related by C^2.

For the rest of the book the Sister 1 solution will be used as the primary solution which relates charge to mass and the constants of the universe. We can then calculate the constants of the universe in terms of each other. Thus repeating equation 2-21:

$$G = 16 \pi e U_0 / (137.036)^3 \qquad \text{met}^2/\text{cou sec} \qquad (2\text{-}31)$$

Equation 2-31 is the best fit for the numerical conversion from the gravitational constant to the electrical permeability constant. This calculates to be:

$$G= 6.67223\text{E-}11 \qquad (2\text{-}32)$$

In standard physics, the relationship between Z_o and h and Q is:

$$Z_0 = 2h/ 137.036 Q^2 \quad = 376.828 \text{ met}^3/\text{cou sec}^2 \qquad (2\text{-}33)$$

From the fine constant relationship in standard physics, we get:

$$hC/[2\pi KQQ] = 137.036 \tag{2-34}$$

The 137.036 comes from the reciprocal of the cosine of 360 degrees over 274 half waves as will be shown in chapter 5, section 5-7. Thus:

$$137/\cos(360/274) = 137/0.999737 = 137.036028 \tag{2-35}$$

The relationship between G and the electrical permitivity constant becomes:

$$G = 16\pi e / [\varepsilon_O C^2 (137.036)^3] \tag{2-36}$$

This calculates to be:

$$G = 6.67224E\text{-}11 \tag{2-37}$$

We can also relate Z_O and GC, thus:

$$Z_O = [GC(137.036)^3] / 16\pi e \tag{2-38}$$

GC calculates to be:

$$GC = 2.000392E\text{-}2 \doteq 1/50 \text{ ohms} \tag{2-39}$$

$$Z_O = 376.751 \text{ met}^3/\text{cou sec}^2 \text{ (ohms)} \tag{2-40}$$

We see that the term GC is an admittance of approximately 1/50 ohms. We can also write an equation for h in terms of GC and Q^2.

$$h = GCQ^2 [(137.036)^4] / 32\pi e \tag{2-41}$$

This calculates to be

$$h = 6.62647E\text{-}34 \tag{2-42}$$

We can also add our standard electrical equation:

$$\varepsilon_O U_O = 1/C^2 \tag{2-43}$$

By feeding the equations into each other using the standard formulas of physics, we can derive all the above equations from the Sister 1 table of units. The net result is that the Sister 1 solution enables us to have a whole set of interlocking equations that relate h, c, Q, Uo, ε_o, Zo, and G. The Sister 1 solution enables all the main constants of the universe to be inter-related. This was not accomplished using the Sister 2 solution or any other solution investigated. This leads one to believe that the Sister 1 solution is the correct solution.

SECTION 2-3: THE LIGHTSPEED EQUATIONS

For the Sister 1 solution we have units of meters cubed over seconds cubed and meters squared over seconds squared and various combinations of powers of meters and seconds. What do these units mean? In this section we will investigate the meaning of such complex units.

When the radius of the universe increases, both the charge and the energy of the universe drop. Looking back in time, we see that as we compress space-time, we charge up the universe and give it energy. We can look at energy from equation 2-22 as charges moving in a space time volume. Thus:

$$\text{Energy (E)} = Q C^3 \tag{2-44}$$

In equation 2-44 we see that energy is charge times the speed of light cubed. In addition:

$$E = M C^2 \tag{2-45}$$

In Equation 2-45 we have Einstein' famous equation. We now see that energy is mass times velocity squared and also from equation 2-44 energy is charge time velocity cubed. Thus solving for M we get:

$$M = QC \tag{2-46}$$

Equation 2-44 and 2-46 are the missing piece of the puzzle that Einstein started to solve. For momentum the equation is:

$$MV = QVC \tag{2-47}$$

Finally for the momentum of a photon, the equation is:

$$M_O C = Q C^2 \tag{2-48}$$

In equations 2-46 we find that mass in coulomb meters per second is a first order linear function of the light speed. In equation 2-48 we see that momentum in coulomb meters squared per seconds squared is a planar function of light speed. Finally in equation 2-44 energy in coulomb meters cubed per seconds cubed is a volumetric function of light speed. These equations permit the reader to see how the entire spectrum of possible universes varies with light speed.

For constant energy solutions, from equation 2-44 we see that charge decreases with the cubic power of light speed. If the operating point of this universe was at 2C or twice our present light speed, the charge of the electron would be 1/8 that of our present charge.

The above equations are light speed C equations. They tell us the properties of possible coexisting universes for a multiple light speed solution. They also tell us what the universe could have been during previous cycles where the light speed could have been much higher. They also help to tell us the condition of the universe right after the big bang and before the little bang.

The above equations are the primary equations of the universe. Einstein was able to present one primary equation (Equation 2-45). This book presents the rest.

SECTION 2-4: THE GRAVITATIONAL CONSTANT

We can now show how the equation for the gravitational constant was derived using electrical theory and the mass to charge conversion charts in section 2-2. This gives us an exact equation that defines the exact relationship between the electrical world and the mechanical world.

We can produce a general electrical equation from magnetic field attraction theory. The exact constant such as π or 4π or no constant depends upon the geometry involved. It is necessary to find the best electrical fit that produces a universe in the order of 15.75 billion years as per the Dot theory and astronomical measurements. The method will then provide us with a second means of deriving the time of the universe since big bang.

Years of studying the constants of the universe by numerical analysis and a simple engineering calculator has demonstrated to the Author that the universe is tied together by the constants 4π, $16\pi e$, 137, and 137.036. These numbers always appear. The inverse fine constant number 137.036 will be explained in Chapter 5. The exact reason for this precise number has been found during these studies.

Numerical analysis shows us that the above factors enter into many equations of modern physics. The Author accepts such corrections as being appropriate.

In general the gravitational force is driven by the discharge of the dot charges throughout the universe. As the universe expands dot current flow of both positive and negative nature extends from all matter and photons outward to the radius of the universe. This causes spherical magnetic fields that are constantly expanding. The space-time memory of the past magnetic field of a few microseconds ago and the present field of right now produces a uniform force. This force can also be viewed as the contracting force acting upon the Bohr Orbit when we view the universe from that perspective. Therefore we can write the gravitational field equations in terms of the expansion of the Bohr orbit.

At the same time as the universe expands, the Bohr orbit expands and this causes a force between the present orbit and the past complete orbit. Likewise a similar force exists between the present proton inner oscillation and the last cycle. They are all Ampere's laws of current loop problems. The force between past and present tends to contain the present from expansion. Thus we have universal gravitation.

The hydrogen atom becomes the standard. The neutron has similar effects and the heavy atoms also have similar effects. Since the Bohr atom is expanding, the protons are also expanding and the neutron radius is expanding as well. The general expansion of the universe common mode produces the decay of charge all over the universe and the general gravitational field.

Let us look at the force between two hydrogen atoms by producing a magnetic attraction equation that relates the spinning electron of one atom interacting with the Bohr Orbit expansion of the second atom. One term will be large, the other small.

$$F = G M_H M_H / R^2 = 2U_O (QC/137.036) [4\pi\ QV_B^*] / (R^2) \qquad (2\text{-}49)$$

Equation 2-49 is a general generic equation relating a gravitational force between two hydrogen atoms to the interaction of the spinning electron in the first Bohr Orbit of one atom producing one magnetic field represented by the term $QC/137.036$, and a magnetic current loop caused by the slow expansion of the Bohr radius of the second atom represented by $4\pi\ QV_B^*$. The factor (2) is caused by the electron of atom 1 reacting with the field of atom 2 and the electron of atom 2 reacting with the field of atom 1. This doubles the force. The velocity V_B^* represents the motion of the Bohr radius as it expands slowly in time. At this time no other corrective constants are used such as $[16\pi e/137.036]$. Thus equation 2-49 represents the exact answer to within a slight correction factor.

The constant G represents the gravitational constant of 6.67260E-11, M_H represents the mass of the hydrogen atom of 1.67353E-27, Uo is the electrical permeability of 1.25664E-6, C is the speed of light of 2.99792E8, Q is the charge of both the electron and the proton of 1.60218E-19, and R is the distance between the two atoms.

The equation for the electrical form of the gravitational constant can be written in two different ways. We can write the equation in terms of the Bohr radius or we can write the equation in terms of the neutron radius. This will be done in this chapter in generic form from the conversion charts. When we look at the neutron, a slightly different equation would apply with different constants since the neutron radius is small and the expansion velocity of this radius is smaller as well.

The Bohr orbit becomes the guide although equations can be written for other things and the electron itself. However, the Bohr orbit equation becomes an easy equation to understand. As the universe expands, both the Bohr orbit's radius and the neutron's radius expand. This enables us to produce gravitational equations in terms of the Bohr orbit expansion velocity or the neutron expansion velocity. We can also write the equations relating to the radius of the universe.

From equation 2-49, we can solve for the velocity of expansion V_B^* of the Bohr orbit.

$$V_B^* = 1.05367E\text{-}28 \text{ meters per second} \tag{2-50}$$

In the Dot theory we normalized the time of the universe by equation 1-39:

$$4\pi Q T_U = 1^* \text{ (coulomb seconds)} \tag{2-51}$$

This gives a time of:

$$T_U = 4.96682E17 \text{ seconds} \tag{2-52}$$

$$T_U = 15.75 \text{ billion years} \tag{2-53}$$

The 15.75 billion years reasonably matches the astronomical data. It is a good yardstick for the Dot theory. Any other means of measuring the time of the universe must match reasonable well. In equation 1-40 we assumed that time was following an exponential curve and that today was the zero time reference point. Thus we got a normalized time of 15.75 billion years

Another measure of the time is to look at the Bohr orbit and to assume that it has been constantly expanding linearly since right after big bang. It too could follow an e^x curve and we could again assume that we are at $e^0 = 1$. Thus t = To = 15.75 billion years.

The standard physics equation for the Bohr radius is:

$$R_{Bohr} = 137.036 \, h / (2\pi M_E C) \tag{2-54}$$

The Bohr radius calculates to be 5.29178E-11 based upon an electron mass of 9.10939E-31 for the Bohr orbit. The time of the universe is the Bohr radius divided by the Bohr velocity. Thus:

$$T_U = R_B / V_B^* = 5.02224E17 \text{ seconds} \tag{2-55}$$

$$T_U = 15.925 \text{ billion years} \tag{2-56}$$

The difference between the Dot theory normalized time and the Bohr atom analysis is:

Difference = 1.109 percent (2-57)

The difference in time could be partly due to the usual correcting constants of $16\pi e$ and 137.036, which may appear twice. It could also be due to the slowing of the universe. In addition, the universe follows an $e^{x \sin t}$ curve rather than an e^x curve. Thus several factors make up the one-percent error. It may very well be that the Bohr orbit equation gives us a means of equating the slowing of the universe to our here and now measurements. This is a topic for future study.

Let us now look at the equivalent equation for the neutron. The neutron's velocity of expansion will be:

$$V_N^* = V_B^* R_N / R_B \quad (2\text{-}58)$$

The neutron will expand at the same rate as the Bohr orbit. The proton will do the same as well as discussed in Chapters 6 & 7 since:

$$R_B / R_N = 2 (137.036)^2 \quad (2\text{-}59)$$

Equation 2-59 is basically identical with the ratio of the Bohr radius to that of the neutron radius. Solving for V_n^* we get:

$$V_N^* = V_B^* / [2(137.036)^2] \quad (2\text{-}60)$$

$$V_N^* = 2.80546\text{E-}33 \quad (2\text{-}61)$$

We can also put equation 2-61 into equation 2-49 and substitute the mass of the neutron for the mass of the hydrogen atom.

$$F = GM_N M_N / R^2 = 16\pi (137.036) U_O Q C Q V_N^* / R^2 \quad (2\text{-}62)$$

In equation 2-62 we can use the mass of the neutron which is larger than the mass of the hydrogen atom. This will cause a slightly higher value for the motion of the neutron radius. In this case V_N^* calculates to be:

$$V_N^* = 2.81016\text{E-}33 \text{ meters per second.} \quad (2\text{-}63)$$

The gravitational constant and the electrical permeability constants were matched by numerical methods. Thus:

$$G = 16 \pi e\, U_O / (137.036)^3 = 6.67223\text{E-}11 \quad (2\text{-}64)$$

In equation 2-64 we see a near perfect match for G to within 0.005 percent of the standard value of 6.6726E-11 by using the general correction factors of $16\pi e$ and 137.036 cubed. Since these factors always appear during numerical analysis, equation 2-64 appears true.

CHAPTER 3 – THE OSCILLATING UNIVERSE

SECTION 3-0 INTRODUCTION

In this chapter the oscillating and expanding universe will be studied from the perspective of an exponential sinusoidal solution and a purely exponential solution.

SECTION 3-1: THE INDUCTIVE/CAPACITIVE OSCILLATOR

In Chapter 1, equation 1-29, the dot energy was calculated to be:

$$E_D = K Q Q / 4 \pi R_U \tag{3-1}$$

where K is Coulomb's constant, Q is the charge of an electron, and R_U is the radius of the universe. We could also represent the energy by using magnetic equations. Thus:

$$E_D = U_O Q^2 C^2 / [(4\pi)^2 R_U] \tag{3-2}$$

In equation 3-2 we see that the dot energy is the electrical permeability constant U_O times the charge Q squared; times the speed of light C squared; divided by (4π) squared; and finally divided by the radius of the universe R_U.

We can also assume that half the dot energy is inductive and half the dot energy is capacitive. Therefore:

$$E_D = (1/2)U_O Q^2 C^2 / [(4\pi)^2 R_U] + (1/2)KQ^2/(4\pi R_U) \tag{3-3}$$

The dot inductance as we look outward from the dot toward the universe can be calculated from the electrical equations of basic physics. Inductance equals the electrical permeability times the area divided by the distance. With a universe of radius R_U and surface area of $4 \pi R_U^2$ we get:

$$L_D = 4 \pi U_O R_U \tag{3-4}$$

The dot inductive energy is:

$$(1/2)E_D = (1/2) L_D (I_D)^2 \tag{3-5}$$

In equation 3-5 we see that the dot inductive energy is equal to the inductance of the dot times the dot current squared. This is the DC dot current which drives the universe. It is not an ordinary current but it is similar to the discharge of a parallel plate capacitor where the plates are moved.

The dot capacitance is equal to the permitivity times the area divided by the distance. Thus:

$$C_D = 4 \pi \varepsilon_O R_U \tag{3-6}$$

The dot capacitive energy becomes:

$$(1/2) E_D = (1/2) C_D (V_D)^2 \tag{3-7}$$

In equation 3-7 the dot capacitive energy equals the dot capacitance, C_D times the dot voltage V_D squared.

The energy of the universe will oscillate from energy stored in the capacitance of space to the energy stored in the inductance of space. The cycle time normalized to our clock is:

$$T_C = 2\pi (L_D C_D)^{1/2} \tag{3-8}$$

In equation 3-8 the cycle time T_C is equal to 2π times the Square root of the dot inductance times the dot capacitance. Substituting for L_D and C_D we get:

$$T_C = 8\pi^2 R_U / C = 8\pi^2 T_U \tag{3-9}$$

With T_U equal to 15.75 billion years, the cycle time is:

$$T_C = 1244 \text{ billion years.} \tag{3-10}$$

This is for an exponential sinusoid function. The universe to us looks like a universe of 15.75 billion years. Beneath the surface there is an inductive capacitive oscillator driving the universe. On the expansion cycle, the magnetic field is slightly stronger than the electric field. The universe then expands. On the contracting cycle the electric field is slightly stronger than the magnetic field and the universe contracts. The inductive/capacitive oscillator is the driving force of the universe. Although the universe follows an exponential sinusoid, the driving function is a perfect sine wave.

From this we see that every dot in space looks like a huge inductor and a huge capacitor. Space itself oscillates since the dot is a focal point of the electromagnetic field. These equations represent the oscillation of space itself of which the dot is a part. Alternately, the dot can be considered an independent space-time bubble like a single hydrogen atom in a general gas distribution. The bubble reaches to the radius of the universe, were all the space time bubbles inter-twine.

You could also consider the dot as a double bubble. It exists in the + Universe as a sphere. Dot current flows into the – Universe through the positive dot which it a pathway between both halves of the universe. At negative dots, current flows the opposite way. To us it looks like one universe. Everything we see is due to the active dot currents at the junction of the universes.

The active dot current universe is like the stuffing in a sandwich. There are two slices of bread but all the action occurs within the bread.

The 1244 billion years is what a variable time clock would see. This time clock keeps slowing with time so it can measure infinite amounts of time with normalized results. That is the beauty of an e^x function. The actual time when measured by a constant frequency non variable time clock would appear nearly infinite. We will look at the solution for the big bang time clock shortly. First we need to study variable space-time mathematics.

SECTION 3-2: VARIABLE SPACE TIME MATHEMATICS

This book is devoted to both an exponential log solution and an exponential sinusoidal solution for the expanding universe. The mathematics of variable space-time is very difficult because the time base keeps changing. A simple differential equation becomes very much more difficult to solve when differential time keeps changing.

The beauty of the exponential solution is that the derivative of e^x with respect to x is e^x. The exponential solution permits us to select a point on the e^x curve and call this time zero. Then we can look forward toward plus infinity and backward toward minus infinity.

The exponential function enables us to mathematically look at infinity while keeping the math simple. The exponential sinusoid has a similar characteristic. In the exponential solution we can call today time zero. Thus for t=0:

$$e^t = 1 \tag{3-11}$$

In equation 3-11 the function e^t for (t=0) is unity. Therefore we can rewrite all the quantities previously analyzed in the exponential form of e^x.

$$M_U = M_{Uo} \, e^{-t/T_u} \tag{3-12}$$

$$T_U = T_{Uo} \, e^{t/T_u} \tag{3-13}$$

$$R_U = R_{Uo} \, e^{t/T_u} \tag{3-14}$$

$$Q = Q_O \, e^{-t/T_u} \tag{3-15}$$

$$K = K_O \, e^{2t/T_u} \tag{3-16}$$

$$U_O = U_{Oo} \, e^{2t/T_u} \tag{3-17}$$

$$\underline{C} = \underline{C}_O \, e^{-t/T_u} \tag{3-18}$$

$$L = L_O \, e^{3t/T_u} \tag{3-19}$$

$$V_D = V_{Do} \, e^{0} \tag{3-20}$$

$$I_D = I_{Do} \, e^{-2t/T_u} \tag{3-21}$$

$$X_L = X_{Lo} \, e^{2t/T_u} \tag{3-22}$$

$$X_{\underline{C}} = X_{\underline{C}o} \, e^{2t/T_u} \tag{3-23}$$

By looking at the chart in section 2-2 for the various items, we can derive these equations. In the MCS system for a constant light speed solution, it is only necessary to see the ratio of the terms. For example if the units are (coulombs/second) then the quantity decreases to a second power with time. If the units are (met³/coulomb second), then the quantity increases with the third power as both time and distance increase and coulombs decrease. If you look at the chart it is only necessary to know that the charge is decreasing over time while both time and distance are increasing. Mass that is coulomb meters per second decreases with the first power of time or distance since both seconds and meters increase and coulombs decrease as the universe expands.

The dot voltage in (meters³/seconds³) is seen to be a constant over the entire cycle time of the universe. Thus it is a constant voltage universe. Since (I = coulombs per second), the universe is a slowly decreasing current source. Plank's constant is a true constant. It will not vary over the entire cycle of the universe. The light speed is constant, however there is a slight hysteresis loop in light speed that permits the universe to take on a sinusoidal oscillation as will be explained later in this chapter. Likewise, there is a slight differential change in meters and seconds, which causes the hysteresis loop.

In these equations (t) is the here and now time that we are used to. This time is slowly increasing, however this increase is at so slow a rate compared to T_U that we could assume T_U is constant. In addition, the exponential is basically

$e^0 = 1$ to a high degree of accuracy and thus we end up with ordinary equations. In order to calculate the dot current that discharges from the dot capacitance we would normally use the expression:

$$I_D = C_D \, d(V_D)/d(t) \tag{3-24}$$

The simple solution would be $C_D \, d(V_D)/d(t)$. The problem is that the dot voltage is always constant. However this expression which has always been common for electrical engineering has always been for circuits in which the capacitance is a constant. However the capacitance of the dot varies. This is similar to the mechanical world problem:

$$d/(dt) \, MV = M \, d(V)/d(t) + V \, d(M)/d(t) \tag{3-25}$$

Likewise:

$$I_D = d/d(t) \, (C_D V) \tag{3-26}$$

The dot current flow is a flow due to a change in dot capacitance. Since capacitance equals ε_0 times area divided by distance we see that as the universe expands the capacitance drops. This is because the electrical permitivity ε_0 drops as a second power of distance as seen from equation 3-16 for coulombs constant K, which varies as the reciprocal of the permitivity ε_0.

The impedance of the universe heads toward infinity as the universe expands as seen by equations 3-22 and 3-23. When we head toward infinity the dot current flow ceases and we have the potential energy of stretched space. All the active energy in the universe is gone at that time. There are no dots; there are neither stars nor planets. There is no life in our universe. Higher light speed universes could possibly exist but our universe is completely dead. At that time it is completely homogeneous. It is a timeless state of pure nothingness.

After an infinity of time, the universe of zero mass and zero kinetic energy will start to compress. Once it is moving, dot current starts to flow. The motion of pure space-time at infinity starts current flow. This will bring us right to the big bang and the creation of our universe.

We then have two solutions. We can oscillate from a minimum radius to infinity and stay at infinity forever. An infinity of time turns forever into a finite time. We know it happened because we are here. It will happen again. The e^x solution is a very simple solution. Infinity becomes the universe. The sinusoidal solution depends upon the light speed hysteresis loop, which will be discussed shortly. For this solution, something must prevent the return to infinity.

The variations of the dot capacitance for a constant dot voltage universe produce the universe that we see. Of course the inductance changes as well.

Taking the derivative of equation 3-26, we get:

$$I_D = C_D \, d(V_D)/d(t) + V_D \, d(C_D)/d(t) \tag{3-27}$$

Since the voltage V_D is a constant:

$$I_D = V_D \, d(C_D)/d(t) \tag{3-28}$$

We can now take the derivative of capacitance Equation 3-18 for the equation of dot capacitance C_D. The solution is $-C_D/T_U$ where the minus sign indicates that the current flow is out of the plus dot. The dot current then becomes:

$$I_D = K Q C_D / R_U T_U \qquad (3\text{-}29)$$

Since $C_D = R_U / 4 \pi K$ we get:

$$I_D = Q / (4 \pi T_U) \qquad (3\text{-}30)$$

As a check we can calculate the dot voltage across the dot inductance:

$$V_D = L_D \, d(I_D)/d(t) + I_D \, d(L_D)/d(t) \qquad (3\text{-}31)$$

In this case since L_D and I_D vary with time, each term contributes to V_D with respect to time. Thus:

$$d(I_D)/d(t) = -2 I_D / T_U \qquad (3\text{-}32)$$

Also:

$$d(L_D)/d(t) = +3 L_D / T_U \qquad (3\text{-}33)$$

Therefore:

$$V_D = -2 L_D I_D / T_U + 3 I_D L_D / T_U = +L_D I_D / T_U \qquad (3\text{-}34)$$

since $L_D = U_O R_U$, $I_D = Q/4\pi T_U$, and $K = U_O C^2/(4 \pi)$

$$V_D = K Q / R_U \qquad (3\text{-}35)$$

The dot voltage produced by the simple method agrees with the dot voltage produced by the more complicated method. This is the beauty of the exponential log.

From this exercise it is understood that the universe operates from the variable inductance and variable capacitance of the dot, and or space. The hysteresis loop will cause it to oscillate as an exponential sinusoid; otherwise it would be a pure exponential log function.

The dot current flow is at the junction between the capacitance of space and the inductance of space. In this regard we have a dual universe. Half the universe is a capacitor and half the universe is an inductor. They are separated in space and time and joined by the dots.

SECTION 3-3: MAGNETIC AND ELECTRIC FIELDS OF THE UNIVERSE

In the Dot theory, there is dot current flow which effect oscillation of the universe over the long period. The dot current flow provides a dot current magnetic field that is counter balanced by the dot electrostatic field. There are forces developed that are equal and opposite and which drive the universe. There is a phase shift between these forces that give rise to a hysteresis loop in space and time. This causes the universe to oscillate as an exponential sinusoid.

If you equally spread all the dots throughout the universe such that it was in perfect balance, you would find a net attraction among all the dots. This is shown in Figure 3-1.

```
+ - + - + - + - + -
- + - + - + - + - +
+ - + - + - + - + -
- + - + - + - + - +
```

Figure 3-1

In the above pattern of dots, we find that the electrostatic attraction is always positive and this would cause the universe to converge toward a pinpoint.

In Chapter 1 section 1-5, we calculated that the force acting on a dot by the model universe is:

$$F_D = K Q Q / 4 \pi R_U^2 \tag{3-36}$$

In equation 3-36 we see that the force acting on a dot is equal to the force of a spherical charge located at the radius of the universe. If the dot is a plus dot, then the universe that the dot sees is a minus spherical dot at a distance R_U away. The total electrical force acting on the universe is the number of dots in the universe N_U times the force per dot. Thus:

$$F_U = N_U K Q Q / 4 \pi R_U^2 \tag{3-37}$$

At any time that the dots are attracting one another in the universe by static coulomb attraction, there must be a repulsive force which is equal and opposite. When we look at the dot current flow in the universe we find that each plus dot sees $N_U/2$ current flows in the opposite direction but only $[(N_U/2)- 1]$ dot current flows in the same direction. Thus:

For a plus dot the current flows seen are:

$$\#I_D (+dots) = [N_U/2] - 1 \tag{3-38}$$

$$\#I_D (-dots) = N_U/2 \tag{3-39}$$

The net current flow for each dot equals the difference

$$\#I_D (+dots) = -1 \tag{3-40}$$

$$\#I_D (-dots) = +1 \tag{3-41}$$

We see from equations 3-40 and 3-41, that each plus dot sees one negative dot current flow, which represents the universe. The minus dot sees one positive dot current flow.

Since each sees the opposite current flow, the electromagnetic forces are repulsive. From electrical theory, the force between two current paths is:

$$F(paths) = (U_O I_A I_B K_A \tag{3-42}$$

where K_A depends upon the geometry of the paths. Assuming that $K_A = 1$ for the model universe, we get:

$$F = U_O I_A I_B \tag{3-43}$$

Where U_O is the electrical permitivity and I_A and I_B both are the dot current flow.

From Chapter 1, Equation 1-25, the dot current flow for the model universe is:

$$I_{Dot} = QC / 4\pi R_U \tag{3-44}$$

We see that in equation 3-44, the dot current flow is the current flow of an expanding sphere of charge Q located at the radius of the universe and moving at the speed of light. This current flows all the way to the dot and through the dot to the other side of the universe and then outward to another sphere of opposite charge. Therefore by putting Equation 3-44 into Equation 3-42, we get:

$$F_{Dot(Magnetic)} = U_O Q^2 C^2 / [(4\pi)^2 R_U^2] \tag{3-45}$$

The total force acting on all the dots is:

$$F_D = N_U U_O [C^2 Q^2] / [4\pi R_U]^2 \tag{3-46}$$

Since $K = U_O C^2 / 4\pi$ we get:

$$F_{Dots(Universe)} = N_U K Q^2 / [4\pi R_U^2] \tag{3-47}$$

Since equations 3-37 and 3-47 are equal:

$$\text{Total Force(electrostatic)} = \text{Total Force (magnetic)} \tag{3-48}$$

From Equation 3-48 we find that the total electrostatic force in the model universe is equal to the total electromagnetic force. Since the Universe contains a slight hysteresis loop, there are slight differences in these forces. When the universe is expanding the decreasing charge drives the dot current flow which causes the electro-magnetic repulsive forces to be greater than the electro-static attractive forces. When the universe is contracting, the charge is increasing and the electrostatic forces are slightly greater that the electro-magnetic forces. The small difference between the two forces is the gravitational force.

SECTION 3-4: MINIMUM RADIUS OF THE UNIVERSE

The electrical solution to the universe calculated the number of neutrons in the universe as per Chapter 1. Thus:

$$N_N = 1.19742E80 \text{ neutrons} \tag{3-49}$$

$$R_U = 1.48901E26 \text{ meters} \tag{3-50}$$

$$R_P = 1.32142E-15 \tag{3-51}$$

$$R_N = 1.31959E-15 \tag{3-52}$$

In equation 3-49 we had previously calculated the number of neutrons to be 1.19742E80 neutrons. In equation 3-50 we restate the radius of the universe taken from the normalized equation of $4\pi Q T_U = 1*$.

Equation 3-51 shows the calculated radius of the proton, and Equation 3-52 shows the calculated radius of the neutron. Both radii are identical with their wavelengths as will be explained in Chapter 4.

As the universe expands, a point will be reached where the electrostatic fields and the magnetic fields balance out and the universe stops. At that time, the gravitational force and binding force on the protons reach zero because the proton inner AC oscillation runs down. The protons will then be destroyed and turned into dots. This is called the little bang. The gravitational constant becomes zero at maximum radius. At that time, the electrostatic forces between all dots in the universe become larger than the magnetic forces and the universe converges toward a minimum radius.

At the minimum radius all the dots will turn into a single ball. The entire universe will become one huge neutron. The big bang is the explosion of a single neutron filled with all the dots of the universe. In order to calculate the minimum radius of the universe we only need to know the radius of the neutron, which is identical with its wavelength, and the number of neutrons in the universe. We then can equate the total volume of the neutrons and the volume of the shrunken universe.

Thus:

$$[(4/3)\pi (R_N)^3]N_N = (4/3)\pi (R_{U(min)})^3 \qquad (3\text{-}53)$$

In equation 3-53, the number of neutrons in the universe is 1.19742E80. Therefore:

$$R_{U(min)} = 6.50411E11 \text{ meters} \qquad (3\text{-}54)$$

Since the present universe is 1.48901E26, the ratio is:

$$\text{Ratio} = 2.16617E14 \qquad (3\text{-}55)$$

The natural log of the ratio is:

$$\ln(\text{ratio}) = 33.009 \qquad (3\text{-}56)$$

Thus the universe cycles from:

$$R_0 e^{-33.039} \text{ to } R_0 e^{+33.039} \qquad (3\text{-}57)$$

Since these numbers are so large, the variation of light speed from minimum to maximum is almost zero. This variation will be calculated in section 3-6. First we must look at the clocks of the universe.

SECTION 3-5: THE CLOCKS OF THE UNIVERSE

In section 3-4 we calculated that the minimum radius of the universe was equal to a full compression of all neutrons. This produces a combined volume of all the neutrons in the universe without any space left over. At full compression, the universe is one molten mess. The laws of physics remain even though the universe is a molten mess. It is merely the laws of compressed light and compressed dots. It is fire and heat unknown to man except that a neutron star explosion will be similar. For the big bang everything in the universe is super hot at the same time.

At the big bang, the universe is 6.50411E11 meters. It is a combined neutron/antineutron bomb. The ratio of the radius of the universe to the radius at big bang is:

$$R_{BB}/R_O = 2.16617E14 \qquad (3\text{-}58)$$

Since the universe follows an e^x function, the actual dimensions of the big bang are approximately:

$$R_{BB} = R_O \, e^{-2.2E14} = 0 \tag{3-59}$$

In equation 3-59 we see that the radius of the universe was zero at the big bang. The previous radius calculated was for our ruler and time clock. Since as we go back in time the ruler shrinks and the time clock speeds up, the actual big bang radius was zero. The electromagnetic field shrunk to a pinpoint. However, the ruler also shrunk to a pinpoint. Therefore we use normalized time. The time of the universe since big bang is 15.75 billion years. This is a normalized time. It is the slope of the e^x curve at the point where (X=0). Thus:

$$T_U = 15.75 \, e^x \text{ billion years} \tag{3-60}$$

When we look backward in time to the big bang, we look at the slope of this curve. It looks like 15.75 billion years ago. However, in reality it was an infinity of time.

Thus:

$$\text{Time since big bang} = \text{infinity} \tag{3-61}$$

The same is true when we look forward to little bang.

$$\text{Time to little bang} = \text{infinity} \tag{3-62}$$

The solution for the LC oscillator as a driving function also used normalized time. It was found that it takes 1244 billion years for the universe to cycle.

$$\text{Cycle Time of the Universe} = 1244 \text{ billion years} \tag{3-63}$$

The actual cycle time based upon actual clock cycles or ticks of a variable speed clock is:

$$\text{Actual Cycle time} = \text{infinity} \tag{3-64}$$

The normalized solution is something we can understand. If the clock remained the same as today and tracked the ruler, then the universe would shrink to 6.50411E11 meters and it would cycle in 1244 billion years.

Twelve hundred forty four billion years is something we can readily understand. In truth it took an infinity amount of time cycles or clock ticks since big bang and will take an infinity amount of time cycles or clock ticks to return again.

At the big bang, the clock is spinning or ticking at nearly infinite ticking speed. At the maximum radius, the clock is almost standing still. We live in between these extremes. There is a window of life in the universe and we exist between extreme heat and cold death. There is a small window of life in an endless universe of infinite time. On the reverse side of the universe, there is no shape, no form. We will go from big bang to life to cold death to form-less-ness. And then the cycle will begin again.

SECTION 3-6: THE LIGHT SPEED HYSTERESIS LOOP

At present we measure a light speed of 2.99792458E8 meters per second. This light speed will vary slightly as we move from the minimum radius to the maximum radius of the universe. The variation of light speed will be extremely small but extremely important. The variation of light speed will produce an oscillating universe from big bang to little bang.

Now let us calculate the variation of light speed from big bang to little bang and the size in time and distance of the hysteresis loop of the universe.

In the exponential solution, the universe oscillates from a minimum radius to a maximum radius according to the e^x function. As calculated in section 3-4, the e^x function varies from e^{-33} to e^{+33}. For this solution we get:

$$R_U = R_O e^x \qquad (3\text{-}65)$$

$$T_U = T_O e^x \qquad (3\text{-}66)$$

$$C_U = (R_O/T_O) \qquad (3\text{-}67)$$

Let us assume that we exist at the zero point where (x=0)

$$R_U = R_O \qquad (3\text{-}68)$$

$$T_U = T_O \qquad (3\text{-}69)$$

$$C_U = C_O \qquad (3\text{-}70)$$

Since the universe is following an exponential function, you can always make today, the zero time. The past will be minus time and the future will be positive time. The mass and energy of the universe can also be expressed as an exponential. Thus:

$$M_U = M_O e^{-x} \qquad (3\text{-}71)$$

$$E_U = (M_O C_O^2) e^{-x} \qquad (3\text{-}72)$$

When the universe expands the mass of the universe drops. The energy drops as well. For the big bang solution, the function (x) varies as follows:

$$-33.009 \leq X \leq 33.009 \qquad (3\text{-}73)$$

The ratio of the mass at big bang compared to little bang is:

$$M_{BB}/M_{LB} = e^{66.018} = 4.6909\text{E}28 \qquad (3\text{-}74)$$

The energy of big bang compared to little bang is:

$$E_{BB}/E_{LB} = e^{66.018} = 4.6909\text{E}28 \qquad (3\text{-}75)$$

In order for the universe to oscillate rather than expand to infinity, a light speed hysteresis loop must exist. The drop in light speed multiplied by the original mass at big bang must equal the residual mass at little bang. Thus for oscillation:

$$(M_{BB} - M_{LB}) \Delta C = M_{LB} C \tag{3-76}$$

Therefore:

$$\Delta C = (M_{LB} / M_{BB}) C = C / 4.6909E28 \tag{3-77}$$

$$\Delta C = 2.13178E\text{-}29 C \tag{3-78}$$

We see that in order for the universe to oscillate, there must be a slight drop in light speed from big bang to little bang so that the loss of momentum caused by the loss in light speed exactly equals the remaining momentum of the universe at the little bang. It is this tiny drop in light speed that prevents the universe from moving outward to infinity.

The result is a light speed hysteresis loop. What does the loop look like? The following diagram illustrates the hysteresis loop.

```
                         x
                  x              x
      BB    x ------------------------------ x    LB        Figure 3-2
                  x              x
                         x
```

In Figure 3-2, the light speed drops from big bang to the present which is a minimum, then it rises upward toward the little bang. The little bang still has a slightly lower light speed than big bang. Thus the line from BB to LB tilts slightly downward.

The net result is that at the present time, the light speed hysteresis loop is approximately twice as large as at little bang. On the way back to big bang, the light speed will take the upper path. The speed of light will rise above the present light speed and fall again at the big bang. Of course there will be non-linear effects as well.

The actual light speed difference today is twice that of equation 3-78. Thus:

$$\Delta C = 4.6909E\text{-}29 C \tag{3-79}$$

This amounts to:

$$\Delta C = 1.4063E\text{-}20 \text{ meters/second} \tag{3-80}$$

The separation distance at the radius of the universe between the present universe and the universe from the previous cycle would be twice the distance of our universe from the straight line from big bang to little bang. Thus:

$$\Delta R_U = 2 \Delta C \times T_U \tag{3-81}$$

Since T = 4.96682E17, we calculate the differential distance as:

$$\Delta R_U = 1.397\text{E-}2 \text{ meters} \tag{3-82}$$

In equation 3-82 we see that the separation in distance between the universe of the far past and the universe of the present is approximately one hundredth of a meter. At the spacing of a neutron, the separation would be:

$$\Delta R_N = \Delta C \, R_N / C = 6.190\text{E-}44 \tag{3-83}$$

The separation time between the universe of the past cycle and the present universe is:

$$\Delta T_U = \Delta R_U / C = 4.6599\text{E-}11 \text{ seconds} \tag{3-84}$$

Finally the separation time between the universe of the past cycle and the present universe at the level of the neutron is:

$$\Delta T_N = \Delta R_N / C = 2.0648\text{E-}52 \tag{3-85}$$

The above calculations define approximately what is necessary for the universe to oscillate. The hysteresis loop is very tiny in both distance and time. Such a loop will prevent the return of the universe to an original state of infinite nothingness. However, the cycle time is still basically infinite.

CHAPTER 4 – PROTONS, NEUTRONS, ELECTRONS, & PHOTONS

SECTION 4.0 INTRODUCTION

In this Chapter, the p* sub-particle will be introduced and many new properties of protons, neutrons, electrons, and photons will be explored.

SECTION 4-1: THE P* SUBPARTICLE

A premise of the Dot theory is that individual dots in free space forms geometric patterns of 137 dots called p* sub-particles. The p* sub-particles contains 68 positive dots and 68 negative dots and an additional dot. The positive p* sub-particle has an extra plus dot, while the negative p* sub-particle has an extra minus dot. The most important property of the p* sub-particle is that it fills the entire universe with an equal distribution of mass/energy everywhere.

From standard physics, the fine constant for the Bohr Orbit is defined as:

$$A = [(2 \pi K Q^2)/ hC] = 137.036 \tag{4-1}$$

It appears that 137 are also the exact number of dots in the smallest Dot theory sub-particle p* which would provide a homogeneous energy distribution throughout the universe. As a check, if we took everything in the universe apart and filled it with p* sub-particles equally distributed throughout the universe, the resulting size of the p* sub-particle would be approximately the same size as a proton or neutron. In Chapter 1, the mass of a dot was defined to be:

$$M_D = K Q^2 / (4 \pi C^2 R_U) = 1.37188\text{E-}72 \tag{4-2}$$

Where K=8.98756E9, Q=1.60218E-19, C-2.99792E8, and R_U=1.48901E26.

The mass of the Model Universe was calculated to be:

$$M_U = C^2 R_U / G = 2.00559\text{E}53 \text{ kg (coulomb meters/sec)} \tag{4-3}$$

Thus:

$$N_U = \text{number of dots in universe} = M_U / M_D = 1.46193\text{E}125 \tag{4-4}$$

The number of p* sub-particles would be:

$$N_{p*} = N_U / 137 = 1.06710\text{E}123 \tag{4-5}$$

The volume of the Universe is

$$V_U = [4 \pi (R_U)^3]/ 3 \tag{4-6}$$

The volume of sub-particle p* is:

$$V_{p*} = [4 \pi (R_{p*})^3]/ 3 \tag{4-7}$$

In addition:

$$N_{p*} V_{p*} = V_U \tag{4-8}$$

In equation 4-8 we say that the volume of p* sub-particles times the number of p* sub-particles in the universe is equal to the volume of the universe. Thus:

$$R_{p*} = R_U / (N_{p*})^{1/3} \tag{4-9}$$

The radius of the smallest sub-particle p* is equal to the radius of the universe R_U divided by the cube root of the number of p* sub-particles in the universe. Since $R_U = 1.48901E26$ and $N_{P*} = 1.06710E123$:

$$R_{P*} = 1.45712E-15 \tag{4-10}$$

In equation 4-10 we see that a sub-particle with 137 dots is about the same size as the proton or neutron. The addition of many sub-particles within the neutron causes it to shrink in size slightly. The neutron is about 1.4E-15 in radius. The exact size will be calculated in this chapter. If we choose a higher number for the number of dots in p* its size would grow even more since there would be fewer sub-particles per universe. Thus it appears that the number 137 represents the smallest number of dots in a sub-particle.

We can also look at the dot as a wave and the p* sub-particle as a 137 phase three-dimensional wave. The basic size of the neutron then becomes identical with an even distribution of wave energy throughout the universe, which is now concentrated within the protons and neutrons of the universe. Let us calculate the number of p* sub-particles per neutron:

As per Equation 1-49, the neutron has:

$$N_N = 1.2209E45 \text{ dots per neutron} \tag{4-11}$$

$$N_{N(P*)} = 8.91168E42 \text{ sub-particles per neutron} \tag{4-12}$$

The neutron has 8.91168E42 different energy levels within it. The proton has slightly less. This show that a homogeneous distribution of energy prior to the big bang was compressed over 8.9E42 times as the universe converged on the big bang during the compression cycle of the universe.

SECTION 4-2: PROTON GENERATION

The universe precipitates protons and neutrons. Protons are the natural structure of the universe since they match an equal distribution of energy and have the identical charge as that of a single dot located at the radius of the universe. Neutrons are approximately the same size as the proton but lack a net charge, which makes them an unstable entity in the universe. The electron is a byproduct of a battle between an equal mass proton/electron swirl of dots. The proton won the battle and the electron lost. The stable configuration of the two is for one to have a large mass and the other to have a small mass.

Throughout the universe, huge protons, the size of the Earth with huge charge and huge mass exist and form the center of galaxies, radio stars, and pulsars. The Dot theory permits very concentrated super protons, super electrons, and super neutrons throughout the universe. Non-linearities in space-time produce a wide variety of interesting things.

The regular proton structure fits into the linear properties of the universe. When the electromagnetic field is compressed to a minimum radius, neutrons, protons, and electrons are precipitated like raindrops from the clouds of dots. The protons and electrons battle each other for the center position and in matter galaxies, the protons win while in antimatter galaxies, the electrons win. Initially protons and electrons are of equal mass but during their production in a swirl of cosmic dots the electron transfers most of its energy to the proton. Even today whenever an electron approaches a proton, a reverse transfer of energy occurs which provides the binding energy of the electron. All of this will be explained in this chapter.

There is a slight hysteresis loop in time and space. The light speed has a slight variation over the cycle time of the universe as previously explained in Chapter 3. At minimum radius, the non-linearities will start to lock the proton into place. There is a balance of forces holding the proton together, yet the initial weak differential force is the force of gravity due to the expansion of the universe. This provides a difference between magnetic attraction and electrostatic repulsion within the proton. The gravitational field is part of this counter-force. However, the motion of the dots within the proton produces additional strong proton self-binding forces as will be explained in Chapter 5. Once the expansion slows sufficiently, the proton frequency runs down, the strong forces will weaken, and the protons will explode. Then the universe will disintegrate into the world of dots from whence it came.

The dot wavelength is:

$$\lambda_D = h / M_D C \tag{4-13}$$

The dot deBroglie wavelength (λ_D) is Plank's constant, h divided by the mass of a dot, M_D and also divided by the speed of light C. Since h=6.62608E-34, M_D= 1.37188E-72, C=2.99792E8, we get:

$$\lambda_D = 1.61109E30 \text{ meters} \tag{4-14}$$

The radius of the universe is:

$$R_U = 1.48901E26 \text{ meters} \tag{4-15}$$

The ratio of the wavelength of the dot to the radius of the universe is:

$$\lambda_D / R_U = 1.08199E4 \tag{4-16}$$

Therefore:

$$\lambda_D = (1096.28 \, \pi^2) \, R_U \tag{4-17}$$

Thus to about 280 parts per million:

$$\lambda_D = (1096 \, \pi^2) R_U \tag{4-18}$$

We see that the dot wavelength to high accuracy is 1096 times π squared times the radius of the universe. In Section 4-1 we calculated the radius of a p* subparticle to be 1.45712E-15. This sub-particle has 137 dots. For a dot to have a wavelength almost ten thousand times the radius of the universe, that dot must travel a huge distance within R_{p*} (radius p*) to get back to its point of origin. In that tiny distance the dot must travel 1096 π² times the radius of the universe. Then it will make the correct geometric pattern.

If we add the energy of two photons together we get:

$$h f_3 = h f_1 + h f_2 \tag{4-19}$$

Therefore:

$$f_3 = f_1 + f_2 \tag{4-20}$$

In equation 4-20 we observe that as we add the energy of one photon hf_1 to the energy of the other photon hf_2 we get a new frequency which is the sum of the two frequencies. Since wavelength is the reciprocal of the frequency times the speed of light, we find that the wavelength will get smaller as we add more dots of energy. Since p* has 137 dots, the corresponding wavelength of p* will be much less.

The geometry of p* must be such that the distance a dot travels from one side of the sphere of radius R_{P*} to the other side must be greater than R_{P*} and less than $2R_{P*}$. Since there are only 137 dots, one could say that the dot travels $2R_{P*}$ per trip across the sphere but $4R_{P*}$ for a round trip back to nearly where one starts from.

From equation 4-18, the total distance per dot for a configuration of 137 dots would be:

$$\text{Round trip distance} = 1096 \, (\pi^2) \, R_U / 137 \tag{4-21}$$

$$\text{Round trip distance} = 8 \, (\pi^2) \, R_U \tag{4-22}$$

To find the number of round trips we divide by the distance $4R_{P*}$ per round trip.

$$\text{Round trips} = 2.01712\text{E}42 \tag{4-23}$$

If we only have a single dot, it would have to travel 137 times this amount to cover the same total distance as the 137 dots. Thus:

$$\text{Round trip for dot} = 2.76346\text{E}44 \tag{4-24}$$

The sub-particle p* is basically a hollow sphere. The proton meanwhile is a densely packed sphere of dots. The average radius for dot travel for a solid is $R_{P*}/2$. In effect the dot in the solid is only going to travel from the outer sphere to the center and back. In the lightly packed p* sub-particle it travels from edge of sphere to edge of sphere. In the proton we see that the dots form a standing wave pattern of distance $R_{P*}/2$. Thus the proton structure causes a different geometry of the dots to occur. The p* dot travels a distance of $2R_{P*}$ each half cycle and a total of $4 R_{P*}$ per round trip. The dot within the proton only travels $R_{P*}/2$ for the half trip and R_{P*} for the whole trip. Thus for the proton:

$$\text{Round trips for dots} = 1.10538\text{E}45 \tag{4-25}$$

Since the proton has 1.21922E45 dots, the geometry does not work out. The proton radius which is the same as its wavelength must be smaller than R_{P*} by the factor of the round trips per dot divided by the number of dots. Therefore:

$$\lambda_P = 1.1054\text{E}45 \times 1.45712\text{E-}15 / 1.2192\text{E}45 \tag{4-26}$$

This calculates to be:

$$\lambda_P = 1.32111\text{E-}15 \tag{4-27}$$

In Appendix 3, the proton mass is listed as:

$$M_P = 1.67262\text{E-}27 \tag{4-28}$$

The proton wavelength is:

$$\lambda_P = h / (M_P \, C) = 1.3214\text{E-}15 \tag{4-29}$$

The proton radius equals the proton wavelength, thus:

$$R_P = 1.32140\text{E-}15 \tag{4-30}$$

We have calculated the proton radius to 0.02% accuracy. The important thing is that the size of the proton and or the neutron is the perfect size in the universe. There is nothing else which fits the equations.

In this section we found that the dimensions of the proton and or the neutron matched the geometric relationships for the size of a particle which when broken apart would fill the universe with a homogeneous distribution of subparticles of 137 dots each. Therefore this universe precipitates protons, neutrons, and electrons when it is compressed to the minimum size calculated in Chapter 3.

SECTION 4-3: THE ELECTRON IN NEUTRON'S ORBIT

In this section let us review the hydrogen atom and then apply the results to the neutron. Within the hydrogen atom, the electron orbits the proton at the Bohr orbit and at a speed of C/137. This is the lowest orbit for the hydrogen atom. The radius of the Bohr orbit is:

$$R_{(Bohr)} = 0.529177E\text{-}10 \text{ meters} \qquad (4\text{-}31)$$

In the hydrogen atom the negatively charged electron moves with a velocity of the speed of light divided by the inverse of the fine constant which is 137.036. If we add additional photon energy to the electron it will move to a higher orbit away from the proton. If it radiates energy from the higher orbit it will move to a lower orbit. Thus the electron captures and absorbs photons and emits photons. This gives us the light spectra.

Let us now look at the neutron. The neutron is merely the lowest state of the hydrogen atom. It is a hydrogen atom that is compressed spherically as with the big bang or a star implosion. In the past, energy was pumped into the neutron. If we take a hydrogen atom and add photon energy, in general we excite the atom to a higher state and free the electron. However, if we spherically pump the energy into the electron, we prevent its escape and force it toward the proton.

The neutron in free space tends to be unstable. A charged particle such as a proton tends to electrically match the structure of the universe that is charged dots of value Q at the radius of the universe. Neutrons do not match the universe. They will tend to break apart and form a swirl of dots. The electron is the mass for the swirl of charges. The neutron is a very simple device. It is an equal amount of plus and minus dots. The dots have both AC motions and DC motions. Patterns of moving dots produce binding energies and magnetic fields. Yet, at a distance the neutron is a perfectly balanced electro-statically but not magnetically.

Dots can have AC oscillatory motion as will be explained in Chapter 5. They can also form patterns of plus dots moving in one direction and minus dots moving in the opposite direction. These will produce magnetic fields perpendicular to the plane of motion. Thus the neutral neutron is not quite neutral.

Let us now investigate the structure of the neutron. Is the neutron a pure collection of dots with inner motions, or is the neutron a proton with a low flying electron in the neutron's orbit? The solution to this problem depends upon the energy calculations for the electron in the neutron orbit. It is possible for the electron to orbit the neutron if the energy of an electron into a low flying orbit is less than the total differential energy between the neutron and the hydrogen atom. The electron in this case would be a line of charge.

However, if the energy required to bring the electron near the proton uses up all the energy difference, then the electron will merely become part of the neutron. In this case, you will not find the electron in the Neutron's orbit.

We can calculate fancy 137-phase spherical waves such that the electron will fit between the radius of the proton and the p* sub-particle radius. However, unfortunately there is still a little energy left over. We do not get a perfect energy match. Now we can start the study.

Let us now calculate the amount of energy released when an orbiting electron from the neutron's orbit moves into a Bohr orbit and the neutron is changed into a hydrogen atom.

Let us first look at a chart of the properties of the neutron, hydrogen atom, proton, and electron.

Structure	Mass in Kg	Mass in MEV
Neutron	1.67493E-27	939.565
Proton	1.67262E-27	938.272
Electron	0.910939E-31	0.510999
Hydrogen Atom	1.67353E-27	938.783

We see that when we move from a neutron to a hydrogen atom we lose 0.785 MEV. (Million electron volts) The Author has modified this chart to six places from U.S. Government standards data. In addition, by adding the electron mass to the proton mass the Author prorated the mass of the Hydrogen atom.

Let us now calculate the energy released when the electron moves from the lowest Neutron orbit to the Bohr orbit. We know the answer is 0.782 MEV. In Section 4-2 we found that the radius of the proton was equal to the wavelength of the proton. Thus:

$$R_P = 1.32142E-15 \tag{4-32}$$

For a homogeneous neutron, the radius of the neutron would be the same as its wavelength. Thus:

$$R_N = 1.31959E-15 \tag{4-33}$$

In equation 4-33 we state that the radius of a homogeneous neutron is identical with its de Broglie wavelength. Thus the neutron is slightly smaller than the proton. For a non-homogeneous neutron, the neutron's radius will lie between the proton radius and the P* sub-particle radius. This is where the high-speed electron would be located. Thus:

$$R_P < R_N < R_{P*} \tag{4-34}$$

In equation 4-34, we see that for a neutron with an electron is its orbit, the neutron radius would lie between the radius of the proton and the radius of the P* sub-particle. Thus:

$$1.32142E-15 < R_N < 1.45712E-15 \tag{4-35}$$

Let us now calculate the velocity of the electron in the neutron's orbit. This velocity will be a steady state velocity for the case where the electron stays in orbit. It will be a final transient velocity for the case where the electron merges into the proton.

The force acting upon the electron from the electrical world is:

$$F = K\ QQ/(R^2) \tag{4-36}$$

This is the standard coulomb equation for the force between the proton and the electron. It is the same as used by Bohr. Let us now look at the centripetal force acting on the electron:

$$F = M\ (V^2)/R \tag{4-37}$$

The opposing force is equal to the mass times the velocity squared divided by the distance. This is the same as Bohr calculated for the hydrogen atom.

Since the velocity of the electron is nearly the speed of light at the surface of the neutron, the mass of the electron is:

$$M = M_E / (1 - (V/C)^2)^{1/2} \tag{4-38}$$

This is Einstein's famous equation for the variation of mass and velocity. This equation must always be added to all Bohr orbit equations. It is necessary even at low speeds as will be seen at the end of this chapter.

Since the forces are equal we can equate both forces to each other.

$$M_E\ (V^2)/\ R\ [1-(V/C)^2]^{1/2} = K\ (Q^2)/R^2 \tag{4-39}$$

Simplifying the equations we get:

$$(V^2)/[1-(V/C)^2]^{1/2} = K\ Q^2/\ M_E\ R \tag{4-40}$$

Equation 4-40 provides us with the relationship between the electron's mass-radius (M_E R) and velocity (V) as it approaches the proton. If we know the velocity, we can solve for the mass-radius. If we know the mass-radius, we can solve for the velocity. However, this solution is more complex and it is best handled by a chart of velocities that match the distance in question.

The problem we have to solve involves the energy difference between the neutron and the hydrogen atom. The differential mass/energy is:

$$\text{Differential Mass} = M_N - (M_P + M_E) \tag{4-41}$$

Since M_N = 939.565MEV, M_P = 938.272MEV, and M_E = 0.510999MEV, we get:

$$\text{Differential Mass/Energy} = 0.782001\text{MEV} \tag{4-42}$$

The Einsteinian mass increase in MEV of Equation 4-42 for the electron must be added to the original rest mass of the electron to produce the total mass.

$$M_{E(Total)} = M_{Eo} + \text{Differential Mass} \tag{4-43}$$

$$M_{E(Total)} = 1.29300\text{MEV} \tag{4-44}$$

The ratio of the total mass of the electron in the Neutron orbit or just before being absorbed by the neutron to the rest mass of the electron is:

$$M_{E(Total)} / M_{Eo} = 2.530338 \tag{4-45}$$

In equation 4-45, we see that the electron in the Neutron orbit or just before absorption has a ratio of the Einsteinian mass to the rest mass of approximately 2.53 times. This means that the electron is traveling at close to 91 percent of light speed. We can now use equation 4-40 to solve for the velocity and the radius. The velocity is easily found from Einstein's formula, equation 4-38. Then we can find the distance from equation 4-40. The alternate way of solution is to prepare a chart. This helps to see what is happening.

We can rearrange equation 4-40 as:

$$R = [K Q^2 / M_E C^2] \cdot [1 - (V/C)^2]^{1/2} / (V/C)^2 \quad (4\text{-}46)$$

The first part of equation 4-46 is:

$$K Q^2 / M_E C^2 = 2.817961E\text{-}15 \quad (4\text{-}47)$$

We can now chart the second part of the equation:

V/C	(V/C)²	$[1-(V/C)^2]^{1/2}$	$[[1-(V/C)^2]^{1/2}]/(V/C)^2$	Radius	Mass
0.919	0.844561	0.3942575	0.4668195	1.31548	2.5364
0.9188	0.844193	0.3947234	0.467575	1.31760	2.53342
0.9186	0.843826	0.3951886	0.4683295	1.3195698	2.53043

In the chart we find that the mass of the electron matches the differential mass of the neutron when the velocity reaches 0.9186 C and the radius is the same as the radius of the neutron's wavelength. Thus the chart tells us that the electron does not maintain an orbit around the proton but merges into it.

The mass that was necessary to merge the hydrogen atom into a neutron is only the Einsteinian mass. The electron is moving at 91.86 percent of light speed as it merges into the proton. Of course, this must be forced by spherical compression. In general ordinary added energy pushes the electron far from the proton.

We see that the neutron is a perfect sphere just like the proton. Although the Author initially believed that a fancy electron in the Neutron's orbit was possible, the final analysis denied that. The neutron is merely a reasonably homogeneous mixture of plus and minus dots. Of course there are AC and DC oscillations within it. Yet, we have a very simple neutron.

SECTION 4-4: ELECTRON DE BROGLIE WAVELENGTH

The de Broglie wavelength is a measure of the energy contained within a free particle or mass. If we look at a light photon we find:

$$\lambda_{Ph} = h/M_O C \quad (4\text{-}48)$$

In equation 4-48 we see that the de Broglie wavelength of a photon with rest mass M_O equals Plank's constant (h) divided by the rest mass M_O and the speed of light C. This is for a free photon. If a mass M_O moves with the velocity V then the corresponding wavelength of the mass is:

$$\lambda_M = h/M_O V \quad (4\text{-}49)$$

In equation 4-49 we see that a mass moving with a velocity V has a similar wavelength to that of a photon. However, they are not exactly the same. As shown previously, the wavelength of a proton is identical with the radius of the proton. Thus:

$$\lambda_P = h/M_P C = 1.32142E\text{-}15 \quad (4\text{-}50)$$

Likewise for the neutron, the wavelength is:

$$\lambda_N = h/M_N C = 1.31959E-15 \qquad (4-51)$$

In this case, the wavelength of the neutron is the same as the radius of the neutron.

The de Broglie wavelength measures the effects of light and also electrons moving in a straight line or curved path. Everything is confined to staying within this radius, R_{P*} in the perpendicular direction of travel. An electron with a de Broglie Wavelength in free space is still confined to R_{P*} in the perpendicular direction. The wavelength of the electron is:

$$\lambda_E = h/M_E V \qquad (4-52)$$

In equation 4-52 we see that the wavelength of an electron depends upon its velocity. The proton does not have this problem because the dots of their existence produce patterns of standing waves that basically start on the outer surface and go to the center and back. The proton is a clear-cut case. The electron is more complex.

When we look at the electron in the Bohr Orbit we do not see a spherical shaped object which can be considered a simple sphere.

The inverse fine constant of prime number 137 indicates that the structure of matter and the electron as well is based upon a 137-phase system. We can picture the electron in the Bohr orbit as a wave with 137 equal components.

The de Broglie wavelength for a free electron moving in a straight line or a simple circle is reduced by the factor of 137 when it behaves like a 137-phase generator. This increases the space the wave can exist in.

SECTION 4-5: THE BOHR ORBIT

Let us calculate the velocity of the electron in the first Bohr orbit state. Bohr did an excellent job of bringing an understanding of the hydrogen atom to man. He compared the light spectra and simple equations to relate the electric field and the centripetal force. He derived a set of equations, which fit into each other. He did not explain the binding energy of the hydrogen atom. He merely used the 13.58 electron volts in his work.

Let us now calculate the Bohr orbit without any need to study the light spectra. Bohr studied the spectra and came to his great conclusions. However this method of empirical analysis which fits the data into the equations lacks the understanding of what is happening in the process. This chapter will show the reader the exact workings of the hydrogen atom and how it was produced from the neutron.

In section 4-3, The Electron in Neutron's Orbit, it was shown that the additional mass of the neutron was due to the electron in a final orbit of radius 1.3197E-15. This produced a gravitational mass of 2.53043 times the rest mass of the electron. The characteristics are as follows:

$$M_E = 2.53043 \, M_{Eo} \qquad (4-53)$$

$$R_E = R_N = 1.31957E-15 \qquad (4-54)$$

$$V_E = 0.9186C \qquad (4-55)$$

In equations 4-53,54,and 55 we see that the mass of the electron in the final neutron orbit is about 2.5 times the rest mass of the electron. The velocity was then calculated using the standard equality between the centripetal force and the coulomb attraction. Thus:

$$KQQ/R_B^2 = M_{Eo}(V^2) / R_B [1-V/C)^2]^{1/2} \tag{4-56}$$

In equation 4-56 we find that the electric force equals the centripetal force when corrected for the gravitational mass increase as per Einstein's formula.

For the electron in the neutron's orbit, the Einsteinian orbital gravitational mass calculated by Equation 4-38 is:

$$M_{Eg} = 2.53043 \; M_E \qquad (1.293 \; \text{MEV}) \tag{4-57}$$

The Einsteinian mass increase from equation 4-57 is the difference between the mass of the electron at the orbital speed and the rest mass of the electron. Thus:

$$\text{Delta Mass} = M_{Eg} - M_{Eo} = M_{Eo} / ([1-(V/C)^2]^{1/2}) - M_{Eo} \tag{4-58}$$

This calculates to be:

$$\text{Delta Mass} = 1.53043 \; M_E \qquad (0.782 \text{MEV}) \tag{4-59}$$

In equation 4-59 the gravitational mass, which gives the neutron extra, weight has a surplus energy of 0.782 MEV.

Together there is sufficient energy to radiate photons and neutrinos and to produce the Bohr orbit. The equations of the Bohr Orbit and the Neutron are complicated by the Einsteinian mass increase. Notice at low speeds such as the first Bohr orbit with V= C/137 and higher orbits, the value of the binding energy will be the same value as the Einsteinian mass increase. Thus we can look at the Einsteinian mass increase and use that for the Binding energy.

Let us now look at the Bohr orbit. The neutron will radiate its surplus energy and the velocity of the electron will move from relativistic speeds to ordinary speeds. Bohr deduced that the first orbit would be at C/137 and R_B at 0.529177E-10. (This is a modern number from U.S. Government data). He showed that there was a constant relationship between the square of the velocity of the orbit and the radius such that:

$$M_E(V_B^2)R_B = \text{Constant} \tag{4-60}$$

For an orbit of the series R_B, $4R_B$, $9R_B$... the velocities would be V_B, $V_B/2$, $V_B/3$. The proton electric field attracts the electron and the electron heads toward the proton. Bohr had no explanation as to why the electron would not be captured by the proton. Yet his basic answers were great.

As the electron moves toward the proton, it's Einsteinian mass increases, which causes additional gravitational repulsive forces away from the proton. In addition, the proton transfers some energy to the electron that increases the mass of the electron and decreases the mass of the proton. This helps stabilize the electron in the lowest Bohr Orbit. Additional magnetic attractive binding forces due to the synchronous motion of the electron's AC field with the proton's AC field accompany this. The fields are not perfect AC fields. They are modulated DC fields that do not form perfect sine waves. Harmonic currents, which are synchronized, form attractive fields. It is complex but the fields are attractive.

Let us now look a little more carefully at the electron in the Bohr orbit at V= C/137 to see if 137 is the correct answer or if some other answer is correct. Let us use Einstein's formula:

$$\text{Correction} = [1-(V/C)^{1/2}] \qquad (4\text{-}61)$$

Let us now look at the differential mass as we change the velocity around C/137. The following chart shows the calculations.

Velocity	Correction	Error	Energy (EV)
C/274	0.9999933	6.56PPM	3.4 EV
C/200	0.9999874	12.5PPM	6.387EV
C/137	0.999732	26.6PPM	13.59EV
C/100	0.999499	50.0PPM	25.55EV
C/10	0.9949	5012PPM	2561EV

From the Chart we see that a velocity of C/10 produces a relativity mass error of 5012 parts per million, (PPM) or 2561 electron volts, (EV). Even at C/274 we get a small error in mass due to Einstein's equation. At C/137 we see that the differential relativistic mass has an error of 26.6 PPM or 13.59 electron volts. Now we know where the binding energy of the hydrogen atom comes from. It is identical to the gain in relativistic gravitational mass as per Einstein's formula. However, it comes from the transfer of mass from the proton to the electron that is identical with the above numbers at the Bohr velocities. The above chart is the proof that the binding energy of the electron is identical to the electrons mass increase with velocity as per Einstein's formula. As the electron moves faster and faster in the Bohr Orbit, it pulls energy from the proton to it.

SECTION 4-6: THE FINE CONSTANT CALCULATION

Let us calculate the fine constant of the Bohr equation and realize what it means.

$$A^* = h\,C / 2\,\pi\,K\,Q^2 = 137.036 \qquad (4\text{-}62)$$

The reason for the 137.036 instead of the number 137 is because the Bohr orbit moves 274 half waves every 360 degrees in one plane and 274 half waves per 360 degrees in the orthogonal plane. These are not pure sine waves but modulated DC or half sine waves. For stability, we must phase lock on one-half wave shape increments as we move from one plane horizontally to the next plane below that.

The angular shift between planes is:

$$\text{Angle} = 360°/274 = 1.31386° \qquad (4\text{-}63)$$

There is a phase shift of 1.3186 degrees between each half wave. The cosine of the angle is:

$$\text{Cosine } 1.3186° = 0.999737 \qquad (4\text{-}64)$$

The inverse of cosine 1.3186° is:

$$1/\text{Cosine } 1.3186° = 1.000263 \qquad (4\text{-}65)$$

The inverse of the fine constant is:

$$137 / \text{Cosine } 1.3186^0 = 137.036 \qquad (4\text{-}66)$$

We see that the fine constant correction factor is due to the fact that each complete Bohr orbit is not exactly the same. They are displaced one-half wave shape from each other. The Bohr Orbit is always rotating. There are exactly 137 cycles in the Bohr orbit but there is a net phase shift between one complete electrical cycle and the next complete cycle.

In standard electrical theory the cosine of the angle always applies to forces or power. In the Bohr orbit there will be an attractive force between every segment of the present orbital configuration and corresponding past and future segments.

CHAPTER 5: AC MODELS OF PARTICLES & PHOTONS

SECTION 5-0: INTRODUCTION

In this chapter, a model of the AC oscillations of the primary particles and the photons will be studied. These oscillations are due to the motion of the dots within the various particles or photons. The motion of the dots within the particles can be considered as a purely AC oscillation of the plus and minus dots which produces a primary frequency and higher harmonics. In addition the dots can spin in a plane or several planes producing a DC magnetic field. In this chapter the AC oscillation will be calculated from the point of view that the majority of the energy of the particle or photon is due to the AC oscillation.

SECTION 5-1: PROTON & NEUTRON AC OSCILLATIONS

In this section, let us understand the nature of the source of energy within the proton and the neutron as well. The basic AC equations for the proton are the same as that for the neutron except for a small difference in the constant. The DC equations for the proton will be different but both will have DC dot current loops and corresponding magnetic fields.

The wavelengths of the proton and neutron are basically identical with their radius because both the proton and the neutron are basically solid balls with internal oscillations. Thus,

$$R_P = \lambda_P = h/(M_P C) = 1.32142\text{E-}15 \tag{5-1}$$

$$R_N = \lambda_N = h/(M_N C) = 1.31959\text{E-}15 \tag{5-2}$$

where h= 6.62608E-34, C= 2.99792E8; Q=1.60218E-19. The oscillating frequencies of the proton and neutron are:

$$f_P = C/\lambda_P = 2.26873\text{E}23 \tag{5-3}$$

$$f_N = C/\lambda_N = 2.27186\text{E}23 \tag{5-4}$$

For a stable condition in space, the inductive reactance and the capacitive reactance of the oscillating proton and or neutron must equal the impedance of space. Thus:

$$X_{Cp} = X_{Lp} = Z_O = 376.731 \tag{5-5}$$

The inductances of the proton and neutron are found by the standard electrical formula:

$$L_P = X_{Lp}/(2\pi f_P) = 2.64283\text{E-}22 \tag{5-6}$$

$$L_N = X_{Ln}/2\pi f_N = 2.63918\text{E-}22 \tag{5-7}$$

The capacitances of the proton and neutron are also found by the formula:

$$C_P = 1/(2\pi X_{Cp} f_P) = 1.86211\text{E-}27 \tag{5-8}$$

$$C_N = 1/(2\pi X_{Cn} f_N) = 1.85955\text{E-}27 \tag{5-9}$$

As a check:

$$f_P = 1/(2\pi (L_P C_P)^{1/2}) = 2.2687E23 \quad (5\text{-}10)$$

$$f_N = 1/(2\pi (L_N C_N)^{1/2}) = 2.27186E23 \quad (5\text{-}11)$$

Now that we know both the inductance and capacitance of the proton and neutron, we can find out the nature of the oscillation within them. According to standard electromagnetic field theory, the self-inductance of a wire loop is:

$$L_W = (U_O /(8\pi \text{ Length})) \quad (5\text{-}12)$$

Solving for the length of the path within the proton, we find:

$$\text{Length} = 8\pi L_P / U_O = 5.28564E\text{-}15 \quad (5\text{-}13)$$

Where $U_O = 1.256641E\text{-}6$. We can then find the ratio of the length to the radius of the proton or neutron:

$$\text{Length}/ R_P = 4.00000000 \quad (5\text{-}14)$$

In equation 5-14 we find that the inductive loop requires a ratio of exactly 4.0000000. Clearly this is not possible if the current loop is oscillating in a circular path around the center of the proton. It is only possible if the current loop moves from the center to the diameter and then back to the center again. This involves a distance of $4 R_P$.

From this analysis we see clearly that the proton oscillation is a spherical oscillation from the center outward to the surface and back again.

Now let us look at the capacitance of the proton and see what it tells us. In standard electromagnetic field theory, the capacitance of two concentric spheres is:

$$C_P = 4\pi \varepsilon_o b(a)/(b-a) \quad (5\text{-}15)$$

In equation 5-15 the distance "b" is the larger spherical distance or outer sphere. The distance "a" is the inner sphere. As "a" gets very small, the capacitance gets very small. If we let $b = R_P$ we find that:

$$4\pi \varepsilon_o R_P\, a/(1-a) = C_P = 1.8621148E\text{-}27 \quad (5\text{-}16)$$

Solving for a we get:

$$a/(1-a) = 0.012665148 \quad (5\text{-}17)$$

$$a = 0.0125066 \quad (5\text{-}18)$$

We see that as far as the capacitance is concerned, the oscillation occurs between $0.0125 R_P$ and R_P. There is a 1.25-percent error between the inductive equation and the capacitive equation. This means that the effective radius of the proton or neutron is slightly larger than that derived from the wavelength. The correction factors appear to be due to geometric considerations. The correction factor ($16\pi e /137$) appears to relate to these geometric differences.

Another consideration is that the expression $E=MC^2$ is an uncorrected expression. Since there are many different types of motions and energies, some correction must be made to that equation as well.

Let us now assume that all the mass and energy comes from the oscillation rather than the spin and magnetic moment, etc. This will give us an idea of the magnitude of the oscillation within the proton and the center of the neutron as well. The object here is to provide a working model of the proton's energy supply.

The peak current within the proton or neutron oscillator is found by equating the inductive energy with the total energy of the proton or neutron.

$$(1/2)L_P (I_P)^2 = M_P C^2 \tag{5-19}$$

$$(1/2)L_N (I_N)^2 = M_N C^2 \tag{5-20}$$

In Equations 5-19 & 20 we equate the total energy of the particles with the same amount of oscillating energy. Solving for I_P and I_N, we get:

$$(I_P)^2 = (2 M_P C^2)/ L_P = 1.13763E12 \tag{5-21}$$

$$(I_N)^2 = (2M_N C^2)/ L_N = 1.14077E12 \tag{5-22}$$

$$I_P = 1.06660E6 \text{ Amperes} \tag{5-23}$$

$$I_N = 1.06807E6 \text{ Amperes} \tag{5-24}$$

This is a very large number. The RMS value of this number is:

$$I_{P(RMS)} = 7.541987E5 \tag{5-25}$$

$$I_{N(RMS)} = 7.5523E5 \tag{5-26}$$

The numbers are too large for a single inductor. The most likely solution is a 137 pair set of inductive capacitive oscillators. The actual equivalent circuit of the proton and neutron is a 137-phase generator.

$$I_P / 137 = 5,5051 \text{ Amperes RMS per phase.} \tag{5-27}$$

$$I_N / 137 = 5,5126 \text{ Amperes RMS per phase} \tag{5-28}$$

Likewise, the AC voltage would be for a single LC circuit:

$$V_P = I_P Z_O = 4.01820E8 \tag{5-29}$$

$$V_N = I_N Z_O = 4.02377E8 \tag{5-30}$$

$$V_{P(RMS)} = 2.84129E8 \tag{5-31}$$

$$V_{N(RMS)} = 2.84521E8 \tag{5-32}$$

$$V_P / 137 = 2.07394E6 \text{ Volts RMS per phase segment} \tag{5-33}$$

$$V_N / 137 = 2.08182E6 \text{ Volts RMS per phase segment} \tag{5-34}$$

In equation 5-33 we see that we have an internal oscillation of about 2 million AC volts within the proton or neutron. The proton has a level of approximately 1 million DC volts at the surface. Therefore these numbers appear reasonable. For the most part this AC oscillation stays within the proton or neutron itself. It shows up as the de Broglie wavelength and as the means by which the gravitational and binding fields work. The AC oscillations of nearby protons and or neutrons will phase-lock with each other and cause them to bind.

The neutron frequencies are slightly higher than the proton frequencies. Neutrons will phase lock with neutrons and protons with protons. A neutron and a proton can phase lock if the neutron loses energy and or the proton gains energy. Thus they must be of identical frequency for proper binding.

We have two options for binding energy. We can bind protons to protons and neutrons to neutrons, each operating at a different frequency. In the deuteron, there is no choice. They can only bind if their AC frequencies are identical. The neutron must lose energy and also share some energy with the proton in the deuteron.

Two protons phase lock readily at zero degrees. Three protons need 60 degrees. Four protons need 90 degrees. Neutrons fit in quite nicely because they can phase lock with the protons but don't have the negative coulomb repulsion forces. Binding forces will be discussed more at the end of this chapter.

The gravitational field can be viewed as the loss of dot charge. This also will reduce the total proton charge Q. The proton AC current and voltage will also drop with time. As the universe expands, the protons run down. The net result is a frequency modulation effect that causes a backpressure upon the protons. This is another way of looking at the gravitational field. It can also be described as a frequency rundown, which results in a backpressure upon the protons and everything else.

We can just look at the cause of gravity as due to Coulomb's law or as due to the AC frequency rundown or red shift of the protons and electrons themselves.

SECTION 5-2: PHOTON'S AC OSCILLATION

In this section, let us look at a model of the oscillation of the photon to determine its AC electrical current and its AC electrical voltage. Let us choose to calculate the Red Hydrogen Fraunhofer line of the solar spectrum. Once the wavelength is known, all the characteristics of the photon in question can be calculated. The method of calculating the one red line applies to all the light spectra and to all the photons from any source. The wavelength of the Red Hydrogen line is:

$$\text{Wavelength Red Fraunhofer Line} = 0.6563 \text{ micron} \quad (5\text{-}35)$$

Since one micron is 1E-6 meters:

$$\lambda_{RED} = 0.6563\text{E-}6 \text{ meters} \quad (5\text{-}36)$$

The frequency of an AC wave corresponding to this wavelength is:

$$f = C/\lambda = 4.5679\text{E}14 \text{ Hertz} \quad (5\text{-}37)$$

The energy of the red photon is:

$$E = hf = hC/\lambda = 3.0267\text{E-}19 \quad (5\text{-}38)$$

The inductive reactance and capacitive reactance of the red photon is:

$$X_{Lp} = X_{Cp} = Z_O = 376.73 \text{ ohms} \quad (5\text{-}39)$$

In order for the photon to remain balanced in space and time, it is necessary that the inductive reactance and the capacitive reactance of the photon match space itself. This is true of all photons for all wavelengths and frequencies. Any mismatch between the impedance of the photon and that of the inductive reactance and capacitive reactance of space itself would destroy the photon.

The inductance of the red photon is:

$$L_P = X_{Lp} / (2 \pi f) = 1.3126E\text{-}13 \qquad (5\text{-}40)$$

The capacitance of the red photon is:

$$\underline{C} = 1/[2 \pi f X_{Cp}] = 9.2485E\text{-}19 \qquad (5\text{-}41)$$

As a check, the frequency of oscillation is:

$$f = 1/[2 \pi (L \underline{C})^{1/2}] = 4.5679E14 \qquad (5\text{-}42)$$

The calculation checks ok and we can now calculate the red photon's AC current flow during the oscillation:

$$E_P = hf = 0.5 L I^2 \qquad (5\text{-}43)$$

Solving for the peak current of the photon, we get:

$$I^2 = 2 E_P / L = 4.611E\text{-}6 \qquad (5\text{-}44)$$

$$I_{Peak} = 2.1475E\text{-}3 \text{ amperes} \qquad (5\text{-}45)$$

$$I = 1.518 \text{ milliamps RMS} \qquad (5\text{-}46)$$

In equation 5-45 we find that the red photon has an oscillating current flow of only 1.5 milli-amperes at the very high frequency of the photon. The corresponding voltage of the photon is:

$$V = I Z_O = 5.7206 \text{ volts RMS} \qquad (5\text{-}47)$$

In equation 5-47 we find that the red photon has a voltage of almost six volts rms at a frequency of 4.6E14 hertz. If we could stop the red photon, we could actually measure this voltage using a very wide bandwidth voltmeter. For a 137-phase photon, the values would be 1/137 times as much.

The AC current will cause red photons of the same frequency to move in harmony with each other. There will be an electrical attraction. The mixing of several photons of nearly the same frequency will produce sums and differences in devices with slight nonlinear characteristics. Such light patterns have already been noticed but it was not realized that the photon was an inductive/capacitive oscillator with an AC current flow.

All the photons can be calculated by this method. As the energy of the photons get stronger, the inductance and capacitance of the circuit will get smaller and it will oscillate at a higher frequency. At the same time both the current and the voltage will increase.

Let us now look inside the photon to see what is happening. The diameter of the ordinary photon that becomes part of the hydrogen atom or other atoms is of the same order of magnitude in radius as the electron, proton, and neutron. The maximum size of the photon should be the universal radius of R_{p*}, the smallest subparticle with 137 dots.

$$\text{Photon radius} = R_{p*} = 1.457E\text{-}15 \qquad (5\text{-}48)$$

The length of the photon will be its wavelength. The red photon of the solar spectra is 0.6563E-6 meters. The length is small but much larger than the radius.

When we get to a neutron its length and radius is the same. More energy shrinks the photon. The photon looks like a little line segment. The sun is sending us a nearly infinite number of little line segments each day. They hit our electrons and protons and are absorbed or reflected. If the photon radius were large, it would just pass through the electrons with no effect. Therefore all the photons must have a small radius. They can be slightly bigger than the proton but not that much bigger. In addition, there can be small variations of the photon radius.

Let us magnify the little photon. In addition let us move along side the photon by bringing our higher light speed camera up to the speed of the photon. If we stand at the center of the photon as we travel at the speed of light, we find that the photon looks like a cylinder. We find that the two ends of the cylinder move toward each other. In addition, we find, that the radius of the cylinder moves from R_{p*} toward zero and outward toward R_{p*}. Yet as we look closely we find that at no time has the whole photon shrunk to zero at any point but that some of the dots are expanding while others are contracting. As a minimum we have a two-phase system. The photon may exhibit multi-phase behavior as well since a multi-phase system has a big advantage of stability.

A simple single-phase photon would tend to break apart. It is very difficult to keep all the dots together if we permit wide swings of groups of dots by themselves. The minute we have a balanced two phase or multiphase system the dot patterns have great strength in keeping together.

The speed of the oscillation of the little tube presents an interesting space-time problem. The tube moves at the speed of light C. The front of the photon tube moves backward relative to the rear of the tube. The front of the tube is moving slower than the speed of light while the rear of the tube is moving faster than the speed of light. We have to know how many trips the front and rear of the tube moves per cycle. Using the inductance formula:

$$L = [U_0 / 8\pi] \times \text{Length} \qquad (5\text{-}49)$$

If we use twice the wavelength in the standard formula for self-inductance, we get:

$$L = 6.5633\text{E-}14 \qquad (5\text{-}50)$$

This is off by a factor of two. This may be due to the splitting of the front to back motion into two separate motions. Half the dots are moving forward and half the dots are moving rearward. This will double the total distance. Likewise if we look at a point at the front, we can picture it traveling rearward and then returning to its origin. This would give it a total distance traveled of 2λ. Then we can move the point forward and then return to the original point. The total distance traveled would be 4λ.

The Author assumes that the measurement of the wavelength of the photons by the various scientists was the full wavelength, and not "null" points, which would produce only half wavelengths. Since the photon was considered a point oscillation, it is possible that scientists have used the null points, which are positive waves followed by negative waves. To make things more complicated, a two-phase or a corresponding 274-phase (2 x 137) photon would tend to confuse the measurements. One would think that the scientists were using the true wavelength. However it is possible that the measurement would only be half the true wavelength. In any event, the true inductance of the photon is calculated using the impedance Z_0 and the frequency.

The capacitance of the photon is similar to a coaxial cable capacitance. This calculation will not be done at this time since the inductive formula gives identical results as the simple electrical circuit formula. The only problem is to understand how many trips the wave made back and forth. Assuming the number of trips was four, let us now calculate the speed of the walls of the photon as they move toward each other.

Although the photon is moving at the speed of light and looks similar to a frequency of 4.568E14, it has also moved four trips along the distance of 0.6563E-6 meters. If we used the normal formulas, it might appear that the photon's walls were traveling at four times the speed of light. Yet, Einstein's clock formula applies. The inner clock traveling at light speed according to Einstein is zero. Actually that is slightly incorrect. The true relationship of the clock traveling at light speed is the number of trips back and forth and the ratio of the number of cycles of the photons oscillation to the measured oscillation of the photon by an observer on the ground. Thus:

$$\text{Velocity} = [4 \text{ trips} \times 1 \text{ cycle per second}] \, C / f_{\text{Photon}} \qquad (5\text{-}51)$$

We see that the actual inner velocity of the photon's motion is the speed of light times one cycle per second divided by the frequency of the photon and multiplied by the number of total trips back and forth. Since the photon frequency is so large, the actual inner velocity is almost zero. The huge frequency has been reduced to a mere 4 cycles per second. If the number of trips back and forth was only two, then the inner frequency is only 2 cycles per second.

Einstein predicted that a clock at the speed of light stands still. In truth, it still ticks. It ticks extremely slowly but is still ticks.

The reader can calculate all the photons by this method. The true nature of the photons is a balanced AC oscillation. This is free running and non-synchronized. It is non-gravitational because the frequency of the photon is not synchronized to matter and the photon does not form a repetitive space-time image. All of matter tends to be synchronized to the various proton fundamental oscillations. The individual electrons are phase-locked to the proton frequency within the Bohr orbits. Within the electron or proton, the photons tend to become phase locked in repetitive orbital patterns and thus possess mass. As the electrons change state however, the photons are not phase-locked and are non-gravitational. Therefore we always have a little more inertial mass than gravitational mass.

Photons also tend to spin and thus have additional field properties. Some of the energy of the photon is in the spin. Yet, for the most part the majority of the energy is in the oscillation itself.

In this chapter we have magnified the photon and looked inside it. It is a simple device with simple characteristics. It is a moving AC field. It is composed of billions of plus and minus dots of very tiny charge. The motion and oscillation of the dots produce a complex multiphase AC field with harmonics. In this chapter we have come to understand that the inner clock of the photon looks like a very slow clock to the photon itself. That is an interesting conclusion to the study of the photon.

As the universe expands and the ruler expands we get a Doppler effect to the far stars, which is negated by the expansion of the ruler. The red shift Doppler is negated by the increased ruler and the Doppler alone would not produce the red shift effect. A second effect is necessary. As the universe expands, both charge and time expands and this causes the photon frequency to drop. Thus the photon AC oscillation slows as the universe expands.

Returning to the calibration equation:

$$4 \pi Q T_U = 1^* \text{ coulomb seconds} \tag{5-52}$$

In equation 5-52 we find that time and charge have a direct relationship. The cycle time or oscillation of the photon, electron, proton, and the Bohr Orbit, will drop as the charge of the dot drops. Every time clock in the universe depends directly upon the value of the charge Q or the little dot charges Q_D.

The red shift is not a Doppler as such. The Doppler is only part of it. The red shift is due to the loss of charge per unit time and or unit distance. Thus we get the red shift law:

RED SHIFT LAW: The Expansion of the universe causes the far stars to move away from us at close to the speed of light. Since the universe expands common mode with the ruler, the far stars are not moving relative to the expanding ruler. They are moving relative to a fixed ruler. Simultaneously the dot charge is dropping within the photon. This causes the photon frequency to drop everywhere. The net result is there are two effects causing the photon to lose energy and one effect causing it to maintain energy. The net result is that the slowing of the photon's internal oscillation due to loss of dot charge insures the red shift.

Of course we could take the opposite viewpoint for the alternate red shift law.

ALTERNATE RED SHIFT LAW: The loss of charge results in a decrease of photon frequency. The expansion of the ruler negates the loss of photon frequency. Therefore the red shift is seen as a Doppler effect with the far stars moving close to lightspeed away from us as the universe expands.

Therefore we have two different ways of looking at the red shift.

SECTION 5-3: ELECTRON'S AC OSCILLATION

Let us now model the electron in the Bohr Orbit in order to calculate the electron's AC oscillation. The kinetic energy of the electron in the Bohr Orbit is:

$$E_{Bohr} = 0.5 Me\ V^2 = 0.5\ M_E (C/137)^2 \tag{5-53}$$

Using M_E= 9.10939E-31, C= 2.99792E8, Q= 1.60218E-19 we get:

$$E_{Bohr} = 2.18102\text{E-}18 \tag{5-54}$$

$$E_{Bohr} = 13.6127 \text{ EV} \tag{5-55}$$

The Red Hydrogen Fraunhofer line of 0.6563 micron has energy of:

$$E_{\text{Red Hydrogen}} = 3.0267\text{E-}19 = 1.8891 EV \tag{5-56}$$

The red hydrogen line comes from the energy difference between the second Bohr orbit and the third Bohr orbit as per the Bohr model. Each orbit has an energy level as per the Bohr relationship:

$$\text{Energy} = 13.6127/n^2 \tag{5-57}$$

$$E_{\text{2nd Bohr}} = 3.40318 \text{ EV} \tag{5-58}$$

$$E_{\text{3rd Bohr}} = 1.51252 \text{ EV} \tag{5-59}$$

The difference between the second and third Bohr orbit is:

$$\text{Delta energy level} = 1.89058 \text{EV} \tag{5-60}$$

The solar red hydrogen line has a slightly less energy level than that calculated from the Bohr model. The gravitational field of the sun will reduce the red hydrogen frequency. In addition, this simple calculation does not include electron spin or magnetic moment effects.

Let us now look at the main energy of the electron. The wavelength of the electron is:

$$\lambda_E = h/M_E C = 2.42631\text{E-}12 \tag{5-61}$$

$$R_{\text{Bohr}} = 5.2918\text{E-}11 \tag{5-62}$$

The ratio of the wavelength of the electron to the Bohr orbit radius is:

$$R_{\text{Bohr}}/\lambda_E = 21.8100 \tag{5-63}$$

The number 8 e is:

$$8\,e = 21.7463 \tag{5-64}$$

When we use correction factors of $(16\,\pi\,e/137)$ and $137/137.036$ we get:

$$21.81\,[(16\,\pi\,e)/137]\,[137/137.036] = 21.7463 \tag{5-65}$$

We see by comparing equations 5-62 with 5-61 times the correction factors, that the general relationship between the radius of the Bohr orbit and that of the electron wavelength is the number 8e.

The physics of the universe always relates things by the numbers, 2, 4, 8, 16, π, e, $16\pi e$, 137, and 137.036. The universe operates as an e^x oscillator but the number (e) has not shown up in the calculations so far. The number exists in the equations but where it comes from has been undiscovered so far. The $16\pi e$ factor is readily recognized as due to some analog function and the 137 is recognized as due to a digital function or quantization.

We see that the number (e) exists in the Bohr orbit. It is the ratio of the radius of the Bohr orbit to the wavelength of the electron. The term $16\,\pi\,e$ is:

$$16\,\pi\,e = \text{Circumference Bohr} / \text{Wavelength of electron} \tag{5-66}$$

The actual number is:

$$\text{Circumference of Bohr}/\text{Wavelength of electron} = 137.036 \tag{5-67}$$

We see that we do not get either a perfect analog solution or a perfect digital solution for the rotation of the electron around the Bohr orbit. If we take the proton as having an oscillation that must phase-lock with the electron in the Bohr orbit, the spin of the proton will change the mesh of the waves. The electron spin will also do the same. Instead of a perfect 137 for a digital solution we would get a different answer because the two oscillations are rotating. In addition as we rotate the oscillations fast enough, their frequency will drop as well.

The energy of the electron is:
$$E_E = M_E C^2 = h\, f_E = 8.18709\text{E-}14 \tag{5-68}$$

The frequency of the electron is:
$$f_E = C/\lambda_E = 1.23559\text{E}20 \tag{5-69}$$

The inductive reactance and the capacitance reactance of the electron is:
$$X_{LE} = X_{CE} = Z_O = 376.731 \tag{5-70}$$

The inductance of the electron is:
$$L_E = X_{LE}/2\pi f_E = 4.85264\text{E-}19 \tag{5-71}$$

The capacitance of the electron is:
$$\underline{C}_E = 1/(2\pi X_{CE} f_E) \tag{5-72}$$

As a check:
$$f_E = 1/[2\pi (L_E\, \underline{C}_E)^{1/2}] = 1.23559\text{E}20 \tag{5-73}$$

The peak AC current within the electron is:
$$I_P = (2E_E/L_E)^{1/2} = 580.88 \text{ amperes} \tag{5-74}$$

The peak AC voltage of the electron is:
$$V_P = Z_O\, I_P = 2.18837\text{E}5 \text{ Volts} \tag{5-75}$$

We see that the electron has a peak AC current of 581 amperes and a peak AC voltage of 219,000 volts. This is all at the super high frequency of 1.236E20 hertz. The measurement of these AC voltages and currents with ordinary instruments would be impossible. However, experiments with light should be able to use the light itself to focus another light beam and also to produce sums and differences of currents. Yet, this most likely has been done to some extent in the study of light.

The 219,000 volts is basically self contained. As you move away from the electron, the voltage is reduced to zero pretty quickly. The same is true of the charge Q which has a DC voltage of a little more than a million volts at the radius of the proton but hardly any voltage a short distance away.

Let us now look at the inductance formula. In standard electromagnetic field theory, the entire formula for the inductance of a coaxial line is:
$$\text{L per unit length} = U_O/8\pi + (U_O/2\pi)\ln(b/a) \tag{5-76}$$

In the prior section on the proton inductance, only the first term was used. There it was explained that 4 trips were necessary to produce the $U_O/2\pi$ term. Therefore:
$$L_P = [U_O/8\pi](4 \text{ trips})\, \lambda_P \tag{5-77}$$

The first term of equation 5-76 is called the self-inductance of the center conductor of the coaxial line. The second term is the inductance between the center conductor and the outer conductor. It may be possible that the first term is not applicable to the proton. Then in order for the second term to equal $U_O/2\pi$ the ratio of b/a must be:

$$b/a = e \tag{5-78}$$

This would say that the oscillation within the proton and the electron as well goes from a maximum of (b) to a minimum of (a) with a ratio (b/a) equal to the natural log e.

The inductance of the proton and the electron holds the energy of the particles. It is easy to see why the correction factors exist, since most of the inductive formulas and capacitive formulas are complex. Now let us calculate the inductance of a circular loop similar to the Bohr Orbit from standard electromagnetic field theory.

$$L = U_O \, a \, (\ln 8a/R_O - 1.75) \tag{5-79}$$

Here a= 5.29E-11, the radius of the Bohr orbit, and R_O is the radius of the electron in the Bohr orbit. In general the electron looks like a cylinder in the Bohr orbit. It has a radius equal or less than R_{P*} which is the general radius of particles and photons.

For this calculation let us use the neutron radius R_N (1.4087E-15) for the radius of the inner conductor of the equivalent coaxial cable model of the Bohr Orbit. The Bohr inductance then becomes:

$$L_{Bohr} = U_O \, R_B \, [\ln 8 \, R_B/R_N - 1.75] \tag{5-80}$$

$$L_{Bohr} = U_O R_B \, [\ln 8 \, (5.2918E\text{-}11 \, / \, 1.4087E\text{-}15] - 1.75 \tag{5-81}$$

$$L_{Bohr} = U_O R_B \, [\ln 3.00521E5 - 1.75] \tag{5-82}$$

$$L_{Bohr} = U_O R_B \, (12.6133 - 1.75) = U_O R_B \, 10.863 \tag{5-83}$$

$$L_{Bohr} = U_O R_B \, 4e \, (1.00093) \tag{5-84}$$

We see that the inductance of the Bohr Orbit is basically:

$$L_{Bohr} = U_O \, R_B \, 4 \, e \tag{5-85}$$

Now we see where the $16 \, \pi \, e$ comes from. It is:

$$16 \, \pi \, e = (4 \, \pi) \times (4e) \tag{5-86}$$

The (4e) factor is the basic term for the inductance of the Bohr Orbit. We now have an equation that produces the (4e) factor, which appears in various equations of the physics of the universe. The factor is due to the inductance of the Bohr orbit and various other inductances.

Let us now calculate the AC current flow within the Bohr Orbit. The electron has an internal oscillation current of 580.88 amperes peak value at 1.23356E20 hertz. As this oscillates at a speed of C/137 around the Bohr Orbit, it will appear as a lesser current and a lesser frequency. The induced frequency will be:

$$\text{Induced Frequency} = f_E/137 = 9.00409E17 \text{ hertz} \tag{5-87}$$

The induced current can be found by comparing the inductive energy with the kinetic energy of the electron. Thus:

$$0.5\ L_{Bohr}\ (I_i)^2 = 0.5\ M_E\ (C/137)^2 \qquad (5\text{-}88)$$

$$(I_i)^2 = M_E C^2 / [137^2\ L_{Bohr}] \qquad (5\text{-}89)$$

$$(I_i)^2 = M_E C^2 / [137^2\ U_O\ R_B\ 4\ e] \qquad (5\text{-}90)$$

$$(I_i)^2 = 6.03282E\text{-}3 \qquad (5\text{-}91)$$

$$I_i = 7.7671E\text{-}2\ \text{Amperes peak} \qquad (5\text{-}92)$$

$$I_i = 5.4921E\text{-}2\ \text{Amperes RMS} \qquad (5\text{-}93)$$

We see that the induced current flowing in the Bohr orbit is about 1/20 of an ampere. Let us now calculate the DC current in the Bohr Orbit.

The charge Q moves at C/137 and the DC current becomes:

$$I_{DC} = (QC/137)/ 2\ \pi\ R_B \qquad (5\text{-}94)$$

$$I_{DC} = 1.0568E\text{-}3\ \text{Amperes} \qquad (5\text{-}95)$$

The DC energy of the Bohr Orbit shares the same inductance as the AC current flow. Thus:

$$E_{DC} = 0.5\ L_{Bohr}\ (I_{DC})^2 \qquad (5\text{-}96)$$

where:

$$L_{Bohr} = U_O R_B\ 4\ e = 7.23049E\text{-}16 \qquad (5\text{-}97)$$

$$E_{DC} = 4.0376E\text{-}22 \qquad (5\text{-}98)$$

The AC energy from equation 5-98 is:

$$E_{AC} = 2.1810E\text{-}18 \qquad (5\text{-}99)$$

The ratio of the DC energy to the AC energy is:

$$E_{DC}/E_{AC} = 1.85125E\text{-}4 \qquad (5\text{-}100)$$

We see that the DC energy within the number one Bohr orbit is about 2 parts in ten thousand of the AC energy. Yet it will have a small effect on the light spectra.

In this chapter we have found where the factor 4e comes from in the various equations. We have come to understand that the Bohr Orbit is similar to an inductive current loop. We find that the DC currents do not produce much energy in the Bohr orbit. We find that the internal oscillation of the electron induces additional current within the Bohr radius. Internally the electron is oscillating at the speed of light. Externally it is moving at C/137. This appears to insure that the induced current within the inductance of the Bohr orbit will have a frequency of $f_E/137$.

In addition to this, the electron will have a spin term. This will cause a DC magnetic moment due to the spin. It will also cause a physical magnetic moment due to the spin. This will be an additional frequency to contend with. The inductance for this will be the inner oscillating loop of the electron of 4.85264E-19 as per equation 5-71. For the spin calculation we can use $0.7071R_N$ as the RMS value of the center of mass and the center of the charge Q. Then we can equate the spin energy to the amount of current required. This is for future study.

SECTION 5-4: ELECTRON FORCES

Let us begin a study of the forces acting on the electron by looking at the equation of the electron in the lowest state of the Bohr Orbit and at the point where it merges with the proton to form the neutron. The equation is:

$$G M_E M_P / R^2 + K Q^2 / R^2 = M_E V^2 / [1-(V/C)^2] R + [U_O(\pi) Q^2 V^2] / 4(\pi^2) R^2 \quad (5\text{-}101)$$

In equation 5-101, the attractive forces between the proton and electron are shown on the left side. The first term is the gravitational attractive force due to the masses of the electron and the proton. This is a very small term. The second term is the electrostatic attractive force between proton and electron. This is a very strong force.

On the right side of the equation is the centrifugal force due to the mass of the electron as it reaches the radius R slightly above the proton. The last term on the right side of equation 5-101 is the electromagnetic repulsion of the electron current loop. This is the self-repulsion of the electron. It is a standard electrical current loop theory equation adapted to the Bohr Orbit. The constant of the loop is such that, a pure electrical charge moving at the speed of light will have the electrostatic and electromagnetic forces equal to each other. Thus:

$$KQQ/R^2 = U_O(\pi)QVQV/4(\pi^2)R^2 \quad (5\text{-}102)$$

When V= C, we get:

$$K = U_O (C^2) / 4\pi \quad (5\text{-}103)$$

Since $K = 1/(4\pi \varepsilon_0)$, and $U_O \varepsilon_0 = 1/C^2$ we find that equation 5-103 is an identity.

We can now reduce the equation for the Bohr orbit to be:

$$KQQ/R^2 = M(V^2) / [1-(V/C)^2] R + U_O (Q^2)(V^2)/[4\pi R^2] \quad (5\text{-}104)$$

Let us now look at the various values of the terms in equation 5-101 to attempt to understand why Bohr and others never paid any attention to the repulsive magnetic force within the atom.

Let us now look at a chart of the forces in the Bohr Orbit at the lowest shell:

Gravitational force = 3.65E-46 (5-105)

Electrostatic Force = 8.27E-8 (5-106)

Centrifugal Force = 8.27E-8 (5-107)

Magnetic Force = 2.204E-12 (5-108)

Ratio Electrostatic to Magnetic = 37,480 (5-109)

From these numbers it becomes obvious that the magnetic forces play only a tiny part in the Bohr equations. As we move toward the proton, the forces increase greatly. When V= 0.9186C the mass ratio is 2.53043 as per Chapter 4, section 4-3. At the surface of the proton's radius of 1.32142E-15, the forces are:

Electrostatic Force = 132.124 (5-110)

Centrifugal Force = 132.293 (5-111)

Magnetic Force = 111.49 (5-112)

We see that at the surface of the proton, the magnetic force becomes quite large. We see that it is the magnetic repulsive force that must be overcome to push the electron into the proton. This force is small at the Bohr orbit but it stops the electron from moving closer to the proton. The gravitational force is always tiny. It is the magnetic force that has been left out of the equations in the past. It is this force which necessitates the binding energy of the neutron. It is this force which springs the electron out from the neutron when the neutron is free of external forces.

SECTION 5-5: PROTON/PROTON BINDING

Let us look at the binding energy of two protons and then extend the principles involved to many protons and neutrons Two protons repel each other from a far distance. When two protons approach each other at close distances in the vicinity of R_p = 1.32142E-15 meters, the repulsive electrostatic forces are overcome by the phase locking of the AC magnetic fields. We now have the law of proton/proton attraction.

LAW OF PROTON/PROTON ATTRACTION: At a distance protons repel each other electro-statically according to the inverse square law of the distance between centers. When protons are close together, protons repel each other electro-statically according to the inverse square law of the distance between centers. At the same time protons attract each other electro-statically and also magnetically due to phase locked AC magnetic fields. The net binding energy or attraction between protons up close is equal to the energy necessary to push two charges together.

To understand this law let us look inside the proton. Within the proton we see billions of billions of little charges in AC and DC type motions. The AC motions give us mass and the DC motions give us the property of mass but also produce magnetic moments, spins, and the like. All the characteristics of particles are due to various AC and DC motions and modulations of these motions. In addition when you move an object you get Doppler effects of all these motions as will be explained starting in the next Chapter.

As we saw at the beginning of this chapter, we have very large AC currents flowing in the proton. We also have the main body charge Q total with a DC rotating current. This is also an AC current, depending upon whether you view the here and now, or the entire cycle.

In any event, the proton is a huge conductor. It is an electrical circuit with positive dot current flows and negative dot current flows. These are actual physical dot current flows, as opposed to the discharge of the dots as the universe expands.

When you bring proton A close to proton B, the proton A's main AC field due to the motion of all the dots encounters proton B's dots which are in constant motion at the same frequency. The two fields tend to phase lock and become quite attractive.

The result of the AC motion of the dots of proton A encountering the dots of proton B and visa versa produces an electrical eddy or induced current in both protons. The negative dots of proton A are shifted toward proton B while the positive dots of proton A are shifted away from the proton B. The same is true for the dots of proton B with respect to proton A. This is similar to an electrical skin effect type force. The net result is a weakening of the coulomb repulsion forces between the two protons. Therefore we have both an electrical magnetic attraction effect and a skin effect binding the protons together. There will be a loss of dot energy and photons will flow out of the combination.

In addition to the AC eddy currents, there will be DC eddy current as well. The DC magnetic eddy currents can be viewed as currents in the plane perpendicular to the line between the two protons. Plus dots circle in one direction and minus dots circle in the opposite direction. The net result is two current loops within each proton, which attract each other. For these loops, the following laws of standard electrical theory, apply:

1) Negative charges flowing in the same direction cause attractive magnetic fields.

2) Positive charges flowing in the same direction cause attractive magnetic fields.

3) Positive charges flowing in opposite directions to negative charges cause attractive magnetic fields.

If we look at the proton itself, and the electron itself within the hydrogen atom, we see that we have many positive dot currents attracting each other, many negative dot currents attracting each other, and finally many positive and negative dot currents attracting each other. In fact there can be little or no repulsive forces at all.

We now have the reason for the internal binding energy that insures the stability of the proton itself. The same is true for the electron itself. The dot current flows within the proton or electron provide totally attractive fields. They become point magnetic fields and are completely attractive.

The proton binding forces are very strong. The DC dot current flows can produce very strong proton binding forces. We have billions of dots within the proton. These dots will flow physically in patterns. Let us look at the following pattern:

```
+ >>>>>>>>>>>>>>>>>>>
- <<<<<<<<<<<<<<<<<<<
+>>>>>>>>>>>>>>>>>>>>
- <<<<<<<<<<<<<<<<<<<
+ >>>>>>>>>>>>>>>>>>>
```

In this dot current pattern of the dots themselves within the proton we find that every current flow is attractive. The positive dot currents attract the positive dot currents, and the negative dot currents attract the negative dot currents. Finally all the positive and negative dot current flows are attractive. We see that within the proton there is a total amount of more positive dots than negative dots but that all the dots can form positive current flows. Even the electrostatic fields when the dots are properly aligned plus/minus/plus, will produce zero net repulsive force or only a slightly repulsive force within the proton.

The same will be true within the electron itself. These dot currents are not the weak gravitational dot currents due to cosmic charge decay with time. They are the very strong fields of moving dots within the electron, which produce fantastic currents. Thus they bind the electron together.

In the future, we will be able to overcome this binding configuration and turn both protons and electrons into pure electro-dot energy. Then we will have cheap and abundant electrical energy and rocket power. This will be discussed later in the space travel chapters.

The internal proton fields serve to bind the proton. In a similar manner, protons are bound to protons with AC and DC eddy currents. There tends to be a phase angle or physical plane associated with this binding. However, it is not possible to keep adding protons together. We now need to produce a model for the binding of protons.

LAW OF VIRTUAL CHARGE: A charge moving around a circle or sphere sees the equivalent of an opposite charge in the center of that sphere.

This law is another way of looking at the binding energy of two protons. If each proton sees an equivalent negative charge in the center of itself, then two protons will see an equivalent double negative charge at the boundary of themselves. In the same manner, three protons will see the equivalent of a charge of 3Q at the center of mass of all of them.

If we take two protons and place them side by side, each will see an equivalent plane of charge equal to twice the charge. The binding energy becomes:

$$\text{Binding Energy} = K(-2Q)(Q)/R_P = 2.18 \text{ MEV} \tag{5-113}$$

In equation 5-113 we find that the binding energy of two protons is equivalent to the energy of the charges Q and minus 2Q at the distance of the proton radius of 1.32142E-15.

It is interesting to look at the various atoms. Since the binding energy is mostly magnetic, equation 5-113 is only an approximation of the forces. If we have two protons the eddy current magnetic fields will have no phase shift problems. The eddy current/dot current flows for two protons can align quite easily. Thus when we look at the motion of the dot charges within two protons, they can be aligned readily.

Once we go to three protons we have problems. We cannot align three protons into perfect alignment. We are forced to go to 60 or 120-degree vector patterns. Four protons would require 90-degree patterns. Neutrons do not add to the total center of charge in equation 5-113. Thus neutrons readily fit into the pattern of the atoms. However excessive amounts of neutrons do not produce stable atoms

In general if we place the total negative equivalent of all the protons in the center of mass of the atom, the far protons will have a very weak binding force. They will have more local repulsive forces and only one central binding force. Therefore eventually the atoms become unstable. Let us now look at some elements:

Lithium will be perfectly stable with three protons in 60-degree synchronization. Yet to balance things better, three neutrons fill in the atom. If we only have two neutrons then we have radioactive lithium. Lithium also works with 4 neutrons but becomes radioactive again with 5 neutrons.

We see that for lithium the magnetic forces prevent the electrostatic forces from destroying the atom. Once we have the basic 60-degree shape, the neutrons can fit in. If we look at helium we start with zero degree phase shift and can add one to two neutron to remain stable. An additional neutron becomes radioactive.

If you look at (Be) you will find four protons. This would work well if we place a neutron in the center. We could then surround the neutron with 4 protons with a 90-degree phase shift. Then we can add another 4 neutrons in between. Thus (Be) with 5 neutrons is quite stable but no other combination of neutrons is stable.

Boron (B) has five protons. This will need a proton in the center instead of the neutron for Be. Stability then will require 6 additional neutrons for the best stability pattern. It does work with 5 neutrons as well.

Carbon with 6 protons will work well if all the protons are 90 degrees apart in all planes. The center of the carbon atom will be empty. If we add 6 neutrons we will have the best pattern.

As the atoms get larger and larger, the electrostatic repulsive forces negate the local binding energy forces between protons. In addition the ability to produce good synchronized angles in large atoms gets less and less. After awhile we can no longer produce stable atoms.

Neutrons have the ability to squeeze together in neutron stars. We cannot produce proton stars. Of course huge protons the size of this Earth can be produced in the big bang. Yet, upon local Earth's, atoms are limited in size.

As the universe expands and the charges decrease, atoms which are stable today will eventually become radioactive tomorrow. In the far future all atoms will be radioactive and disintegrate. Yet in the far past, some atoms existed which perished long ago due to radioactivity.

Basically radioactivity occurs when the magnetic binding forces of both DC and AC fields cannot hold the atom together against the proton static fields and the motion of the potential electrons within the neutrons. Some atoms are slightly in balance and a little heat or mere probability of motion will cause protons and electrons to be ejected. Instability will cause the neutrons too to become dislodged.

There is one other law of interest when groups of dots are forming protons and electrons but which also applies to nuclear structure.

LAW OF CHARGE PRODUCTION: Neutral bodies of dots which revolve around each other will trade charges with each other to maintain the integrity of the package. The electron will lose an equal amount of plus and minus dots to the proton during the big bang phase. At the same time, the electron will give the proton more plus dots. This will make the electron negative and the proton positive. The resulting electric field then prevents the loss of the total body and the hydrogen atom is born.

This same law applies to the binding of protons, chemical bonds, etc. A certain amount of plus and minus dots will move between protons to balance out the forces.

The amounts of dots per proton and electron are only a statistical average. Some will have more dots, some less. A certain amount of dots are required for the main structure of each particle. Once we add photons, and motions, we produce a spectrum of dots for each electron. Thus no two electrons are absolutely identical in the universe.

Patterns of dots at the speed of light attempting to circle each other will find that the electrostatic and magnetic forces are equal and opposite. This was shown in the equations of section 3-3. The forces on groups of dots tend to be of equal and opposite nature as the electromagnetic force negates the electrostatic force. In addition, the gravitational centripetal force is also negated by the gravitational centrifugal force. Thus a delicate balance of forces occurs.

However, galaxies do form and the universe expands and contracts. There is a light speed hysteresis loop in which groups of dots attract each other slightly from big bang to little bang. Simultaneously the inverted effect causes the universe to be driven into expansion. From the little bang to the big bang, groups of dots have a slight net repulsive force while the universe is driven to compression. This causes the contracting universe to be homogeneous.

During the little bang the magnetic forces and the electrostatic forces become exactly equal at the crossover point and gravity is zero. This causes everything to self-destruct. The universe suffers two major events each cycle. There is an explosion at big bang and an implosion at little bang. An implosion may be a better word for the little bang since the universe is set upon the course of contraction. At big bang, gravity is also zero for the split second. That is another reason why the universe does not sit in the condition of a molten mess for long.

CHAPTER 6 GG/MM/ DOPPLER LINEAR SPACE TIME

SECTION 6-0: INTRODUCTION

In this chapter we will look at the original work of Lorentz/Einstein and add to it Doppler Radar principles together with the classical linear space-time interpretation of the Michelson/Morley experiment. Einstein produced an orbital space-time interpretation of the Michelson/Morley experiment. This is a steady state interpretation involving cyclical motion. Basically Einstein's solution is a steady state electrical circuit solution. It is excellent for orbital motion and clocks moving around the Earth. It is not good for linear transient motion, which is defined by classical physics.

Linear space-time is important for future high-speed space travel. It is linear classical physics that determines the amount of energy required for traveling to nearby stars. Einsteinian orbital space-time works well in describing the Bohr orbit, the planetary motions, the clock in motion around the Earth, etc. It is not applicable for linear motion moving toward the speed of light. It limits man's ability to understand and achieve high-speed space travel.

In this chapter, we will look at the space-time solution from a general single frequency describing function method in which the Y and Z dimensions are invariant. We will not include non-linear analysis, Fourier series, or spectrum analysis at this level of the work. In Chapter 7 we will show how to derive Doppler Orbital space-time from classical principles.

SECTION 6-1: GG/MM/DOPPLER SPACE TIME PRINCIPLES

The moving electromagnetic fields of an object produce inertia, kinetic energy, and all the other properties of the mechanical world. GG/MM/ Doppler space-time shows what the mass of an object looks like in the forward direction and also in the reverse direction. It is the complete space-time solution as opposed to Einsteinian space-time, which is a reasonably valid but partial solution.

The big problem with the Einsteinian space-time solution is that it applies steady state electrical theory as the general solution to space and time while ignoring transient electrical theory. In general the steady state solution is a special case of the transient solution. Einsteinian space-time would lead one to believe that the experiments in the cyclotron for orbital motion are valid for linear space travel. He ignored standard classical physics and applied orbital or steady state theory to linear motion. This produced the clock paradox and other misunderstandings of the true nature of space and time.

We know that light is mass-less in the forward or rearward direction. We know that light bends around stars and thus has mass in the perpendicular direction. Einsteinian space-time does not take into account the differences in mass in the X, Y, and Z directions. GG/MM/Doppler Space time does.

Let us look at a perfectly stationary sphere in pure outer space that has a gravitational field extending evenly in all directions. We can look at the sum total of all the inner fields and motions of all the electrons, protons, and neutrons as equivalent to a single AC gravitational field located at the center of mass of the object. The minute we move this sphere, the gravitational magnetic field lines in front of the sphere become compressed while the gravitational magnetic field lines in back of the sphere become elongated. Of course every Bohr orbit and every proton, electron, and neutron experiences this effect

As you look at the Bohr Orbits, the front to back oscillation in the direction of motion starts to turn perpendicular to the direction of motion. The size of the X direction decreases while the size of the Y and Z direction increases. As more photons are added and the velocity increases, the Bohr Orbits and the individual neutron, electron, and proton oscillations flatten in the X direction of motion. As we move toward higher speeds, the X direction shrinks rapidly. At very high speeds, a sphere becomes a flat penny.

In this chapter we will consider only the shrinkage of the X direction. The Y and Z directions can remain the same size, they can shrink, and they can expand. All this will depend upon various constraints. In general if you compress an ordinary object in the X direction, the Y and Z dimensions will expand and the volume of material will attempt to remain constant. In this chapter we will look at the constant Y/constant Z solution for variable X. This will match the Einsteinian solution and it will enable the reader to see the basic error of Einstein in his space-time formulas.

If we keep adding photons to the ball, the front gravitational magnetic field lines will be crushed together while the back gravitational magnetic field lines will form a tail. The result will be a Fourier series complex gravitational field. An ordinary block of iron is a complex spectrum of Fourier series gravitational fields. This is an extremely complex electromagnetic problem. However, complex things are often treated with simplified describing functions or equations, which reasonably explain the phenomenon.

Let us now write equations for the spherical object (ball of mass) which stands stationary. Let us assume we are deep in space away from any strong gravitational fields and clear of any rotation from the galaxies since all these produce errors. The ball can be represented as a total mass at a particular wavelength.

$$\lambda = C/f = h/MC \qquad (6\text{-}1)$$

In equation 6-1, the wavelength of the object at rest is equal to the speed of light divided by its frequency. This is also equal to Plank's constant h divided by the mass of the object and the speed of light. For example the wavelength of a neutron is 1.31959E-15. The corresponding frequency of the neutron is:

$$f_N = C/\lambda_N = M_N C^2 / h \qquad (6\text{-}2)$$

The frequency of the neutron is the mass of the neutron times the speed of light squared divided by Plank's constant h.

$$f_N = 2.27186\text{E}23 \qquad (6\text{-}3)$$

where M_N=1.67493E-27, C=2.99792E8, and h=6.62608E-34. The frequency of the neutron is very high. From Equation 6-2 we see that the equivalent frequency of a large ball would be huge. However, this is not true. The frequency of a large ball is the same as the neutron but the amplitude of the gravitational field is increased. The gravitational field of a huge ball is equal to the equivalent number of neutrons times the field strength of one neutron. Thus:

$$\text{Ball's Field} = N_N \times \text{Neutron Field} \qquad (6\text{-}4)$$

Let us now move the ball with a velocity V_B. If we place an observer in back of the ball and another in front of the ball, and measure the mass frequency of the ball, we get for a ball moving toward the right:

$$f_{ML} = [C/(C+V_B)] f_O \tag{6-5}$$

$$f_{MR} = [C/C-V_B] f_O \tag{6-6}$$

In equations 6-5 and 6-6 we see that the Doppler Mass frequency of the ball is smaller to the left behind the ball and larger to the right in front of the ball. Although Einstein had a simple singular total mass, the reality is that a simple Einsteinian type mass does occur in the Y and Z directions but a dual Doppler mass also occurs in the X direction. The equations for the Doppler masses are similar to radar frequencies. However, the Doppler mass equations have a constant center frequency that is invariant while the Doppler radar transmitter frequency varies with velocity as per Einstein's time equations. For Doppler radar we get two frequency variations. For a transmitter upon a plane we get:

$$f_L = [C/(C+V)] f_O [1-(V/C)^2]^{1/2} \tag{6-7}$$

$$f_R = [C/(C-V)] f_O [1-(V/C)^2]^{1/2} \tag{6-8}$$

In Equations 6-7 and 6-8, the center frequency (f_O) aboard the aircraft is slowed by Einstein's formula $[1-(V/C)^2]^{1/2}$. The frequency to the right is the combination of the Einsteinian decrease in airplane transmitter clock and the Doppler increase $[C/(C-V)]$. For the left frequency, we have both an Einsteinian clock speed drop, and a Doppler drop $[C/(C+V)]$.

The Doppler mass is slightly different. As we speed an object up, the gravitational mass in the X direction becomes more inertial and less gravitational. It is the reverse of what happens to the photon as we slow the photon. Thus:

$$M_{Xg} = M_O [1-(V/C)^2]^{1/2} \tag{6-9}$$

In equation 6-9 we see that an object with rest mass of Mo has a decreasing gravitational mass in the X direction as the object speeds up. Thus less and less energy is required to bring an object to higher and higher speeds as we move in free space. This is due to linear space-time theory. It is not what happens in the cyclotron, where orbital steady state space-time theory applies. The gravitational masses in the Y and Z directions for an object moving in the X direction in free space are:

$$M_{Yg} = M_{Zg} = M_{Xg} / [1-(V/C)^2]^{1/2} \tag{6-10}$$

In equation 6-10 we see that the mass of an object in the Y or Z direction which is moving in free space in the X direction is larger than the mass of the object in the X direction by the Einsteinian formula. We see from equation 6-9 that the mass of the object in the X direction drops with velocity. Thus the total mass of an object in the Y or Z direction is:

$$M_{Yg} = M_{Zg} = M_O \tag{6-11}$$

In equation 6-11, which is for the constant Y/constant Z solution, we see that the mass of an object in free space in the Y or Z directions is invariant. Let us now return to the X direction. In that direction, the mass has dropped. However we have the Doppler masses to consider. Thus:

$$M_L = [C/(C+V_B)] M_{Xg} \tag{6-12}$$

$$M_R = [C/(C-V_B)] M_{Xg} \tag{6-13}$$

We see in Equation 6-12 that the Doppler or inertial mass to the left is smaller than the gravitational mass. The gravitational mass is decreasing with velocity and the Doppler mass to the left is also decreasing with velocity. In Equation 6-13 we see that the Doppler mass to the right is increasing and this is stronger than the drop in the gravitational mass. The total mass is the inertial mass in the X direction. It is a combination of gravitational mass and photon energy.

The solution for the Doppler mass is a complex spectrum of frequencies and harmonics. If you add up all the neutrons, protons, and electrons, you get a spectrum of frequencies and a Fourier series. The describing function approach produces a single frequency equation that gives the salient points of the solution. The inertial mass in the X direction is the root mean square of the Doppler masses. Thus:

$$M_{Xi} = M_{Xg} / [1-(V/C)^2]^{1/2} \tag{6-14}$$

Substituting Equation 6-9 for M_{Xg} we get:

$$M_{Xi} = M_O \tag{6-15}$$

We see that the inertial mass in the X direction is invariant from a single frequency describing function solution. As we move toward light speed we will get some distortions. However, we can now solve the basic problem of the energy required to bring an object up to light speed. Since the mass is invariant in all directions but keeps changing from gravitational mass to inertial mass in the X direction, the solution is similar to classical physics.

$$E = M_O C^2 + \tfrac{1}{2} M_O C^2 \tag{6-16}$$

We see that it takes approximately half the rest mass in energy to bring an object up to near lightspeed. Once we almost achieve light speed, the distortions are a great problem. Additional energy equal to $\tfrac{1}{2}M_O c^2$ is necessary to account for the harmonics. For space travel, a trade off is necessary for safety verses time and distance since at present we cannot survive the distortions.

In orbital space-time, the mass keeps doing the same circular track over and over again. This builds up an image of the object. In the case of the cyclotron we keep pumping energy into the machine and forcing the electron or proton to move in circles. We pump up the mass and produce a huge line of charge. Einsteinian space-time is also true for the Bohr orbit. Thus for purely orbital problems:

$$M_{Yg} = M_{Xi} / [1/(V/C)^2]^{1/2} \tag{6-17}$$

In equation 6-17 we see that the gravitational mass in the Y direction is equal to the Einsteinian formula based upon the total or inertial mass in the X direction. Thus:

$$M_{Yg} = M_O / [1-(V/C)^2]^{1/2} \tag{6-18}$$

We see that for orbital motion the gravitational mass becomes quite large when we approach the speed of light because the perpendicular mass resists the turning. In linear motion, the X masses resist the motion and these eventually turn into pure photon energy, which no longer resists.

The photon has zero gravitational mass in the X direction. It is very soft. In the perpendicular direction, it has mass and is relatively hard. It is hard to bend a light beam. However, a light beam does possess mass and does bend under the gravitational field of a sun. The photon is the purest case. The masses are:

$$M_{Xi} = M_O \tag{6-19}$$

The photon has all inertial mass in the X direction. This makes it soft. The gravitational mass in the X direction is:

$$M_{Xg} = 0 \tag{6-20}$$

We see that the photon has no gravitational mass in the X direction. The gravitational mass of the photon in both the Y and Z directions is:

$$M_{Yg} = M_{Zg} = M_O \tag{6-21}$$

The energy of the photon is:

$$E = M_O C^2 \tag{6-22}$$

The momentum of the photon in the X direction is:

$$P = M_O C \tag{6-23}$$

We see that when we bring a mass up to light speed in linear space-time, it becomes light. Light speed travel for future man has a lot of obstacles to overcome. If we could bring a spaceship up to the speed of light, it would become light itself. In general this would cause it to disintegrate unless future man can devise protective fields to prevent self-destruction.

SECTION 6-2: LINEAR SPACE-TIME

Let us now start the analysis for the properties of linear space-time. The great work of Einstein advanced the thinking of man into the realm of space and time. However there was a lack of understanding of the differences between gravitational mass, inertial mass, and Doppler mass. The problem lies in a restricted singular interpretation of the Michelson/Morley experiment. This interpretation works well for steady state orbital motion but is no good for ordinary classical linear motion.

Now let us look at the Michelson/Morley Experiment again. From the work of Lorentz and Einstein, two postulates were specified:

Postulate 1, the Principle of Relativity stated that the laws of physics are the same or invariant in all inertial systems. The mathematical form of the physical laws remains the same. (Einstein)

Postulate 1 by Einstein is basically true for reference platforms moving at low velocities compared to the speed of light. However in general it is invalid. We live in a nonlinear universe and the laws of nonlinear physics apply. Our Earth moves slowly so the simple describing functions approximate the truth. If we move at higher velocity, the higher harmonics of the nonlinear expressions take over more strongly. At low speeds most harmonics will exist but at higher speeds some harmonics will become as strong as the fundamental. For example, the first Bohr orbit produces a fundamental frequency but high velocity turns this into a Fourier series. Since there are billions of atoms in any mass, we get a spectrum of Fourier series.

The equations of motion and the relationship between large objects moving at very high speeds with respect to each other becomes very difficult to predict. All we can really do is define the motion of high-speed objects with respect to the Earth since it moves slowly in the universe.

Einstein looked at the universe from a mathematical perspective. He took a geometric mean approach. This is pretty good since the Earth is moving very slow compared to the speed of light. His basic answers are true for many areas of space. He felt that every place in the universe was the same but that is not true. There are many places that are quite different than we know. There are many non-linear areas of space-time where human life is not possible.

When two objects are moving toward each other at high speeds, nonlinear electrical theory applies. Objects moving near light speed have different relationships with each other than objects moving slowly compared to the speed of light or objects moving at high speed with respect to an Earth.

There is a big difference between an object in pure free space under the influence of its own tiny gravitational field and an object on planet Earth with the Earth's gravitational field. The laws of pure free space are more like classical physics than Einsteinian relativity. In spite of this, Einstein got excellent results for orbital problems. The main concern of this book is high-speed space travel to our sister planets and nearby stars. We want to achieve at least 50% of light speed. Linear space-time is the most important part of general space-time of which Einstein's orbital space-time is only one particular solution.

Postulate 2 by Lorentz/Einstein stated: The speed of light in vacuum is a constant, independent of the inertial system, the source, and the observer.

Postulate 2 is only partially true. Constant velocity light speed is a very simple solution. It is true that the ideal light speed in pure free space is basically constant over the entire cycle time of the universe. However, this is light speed and not photon speed. The photon speed depends upon the media. It depends upon the motion of the receiver and the sender of the photons. It depends upon the strength of the gravitational fields. It depends upon whether the gravitational fields are moving or relatively stationary. The speed of light is quite variable.

In general, every cubic meter of vacuum in the universe is different than every other cubic meter. The intensity and direction vectors of the electric, magnetic, and gravitational fields in every cubic meter of space in the universe are different. In addition, every cubic meter of space contains an image of the entire universe. The net result is that the light speed in vacuum is different everywhere in the universe.

In Section 6-1, we saw that the mass of the photon in the X direction becomes gravitational mass as it goes from the speed of light to basically zero linear speed within matter. We can calculate the amount of photon energy per unit mass verses velocity. In general, if we took a single photon and reduced its speed in a dielectric, it would become part photon/part mass. The relationship will vary with the velocity squared as compared to the speed of light squared. The relationship was previously stated in Equations 6-14 & 15. Rearranging the terms we get:

$$M_{Xg} = M_O \left[1-(V/C)^2\right]^{1/2} \qquad (6-24)$$

$$M_{Yg} = M_{Zg} = M_O \qquad (6\text{-}25)$$

In Equation 6-24, the gravitational mass of the photon in the X direction varies with its velocity. When it is moving at the speed of light, it has only inertial mass but no gravitational mass. It appears mass-less in free space in the X direction. It still has mass in the Y and Z direction as shown by equation 6-25. Stars will attract it.

The photon is an independent inertial guidance system. Without any forces acting upon it, it will travel at the speed of light. Once you add the synchronized AC fields within a dielectric to the photon, it will slow. Likewise the strong gravitational field of a star will slow the light as U_O and ε_O both increase near the star and it will bend more readily around the star. It bends because it is pure gravitational mass in the Y and Z direction and becomes part mass/part photon in the X direction as it drops speed.

Let us now look at what is happening in the Michelson/Morley experiment. A test instrument is built with perpendicular arms. Light from the sun enters the instrument and a comparison is made of the travel time of the light beam in both directions. A null results and equations are produced for the instrument.

The motion of the galaxy is common to everything, so this experiment will be independent of the galaxy motion. The same is true of the motion of the solar system within the galaxy. The Sun/Galaxy motion will not appear readily using this test instrument. More sophisticated instruments can be readily built which will pick up the absolute motion of the Earth including the galaxy motion. However for the moment we have a simple instrument independent of Sun and Galaxy motion.

The instrument was calibrated and one arm was faced toward the sun when the Earth was moving fastest in its orbit. The instrument nulled. It didn't matter if a light on a bench or the sun itself was used. It didn't matter if the Earth was moving at maximum speed toward or away from the Sun. The null was independent of the relative motion of the Earth and Sun. The speed of light was thought to be constant. It was felt that all you get is a Doppler red shift or a Doppler blue shift when moving away from the sun or toward the sun respectfully. Einstein then proceeded with the two postulates to produce special relativity, which is acceptable for an orbital solution but quite wrong for linear space-time.

The big problem is that we think in terms of infinite light speed and Einstein and Lorentz made the world quite limited with our light speed as the only possibility. They turned three-dimensional space-time into a simple electrical circuit on a bench. They produced an electrical universe similar to a bench circuit. They didn't account for a possible spectrum of multiple light speed coexisting universes, or a universe which may presently operate at our light speed but which can go from zero light speed to infinity light speed over an infinity of cycles and configurations.

The universe we live in has infinite light speed capability. Simultaneous events at near infinite light speed can and do occur all over the universe. Einstein and Lorentz incorrectly omitted the entire light speed spectrum. If we move toward infinite light speed in our minds, then we can look at what is happening in the instrument without limiting ourselves the way Einstein and Lorentz did.

As we look at the instrument, we find a photon coming from the sun moving at its normal self propelled inertial speed of C. In pure free space this will be at maximum ideal light speed. Through gravitational fields, it will be slightly less than the perfect ideal light speed. This photon enters the Earth's atmosphere and slows slightly due to the air molecules that reduce the electrical permeability/permitivity constants. There is also a small effect due to the interaction of the gravitational field of the Earth. The photon will start to synchronize with the Earth and drop its speed slightly as well. In addition there will be an even smaller effect due to a Doppler gravitational field effect. Nevertheless, the Earth is moving toward the photon with a basic speed of V_E. The differential velocity as viewed from an infinity light-speed reference plane between the Earth and the photon is:

$$\text{Differential Velocity} = C + V_E \tag{6-26}$$

Equation 6-26 is what we would see in classical physics. It is also what we would see from an infinite light-speed reference plane. When the photon enters the test instrument, it is still moving at the same differential velocity that equation 6-26 indicate. However, the internal AC gravitational field spectrum of the instrument will cause the photon to adapt to it. The photon will develop a mass in the direction of travel with respect to the test instrument and the speed will be less than shown above. In spite of this, the photon will travel to the back of the test instrument and re-bound forward with a differential velocity:

$$\text{Differential Velocity} = C - V_E \tag{6-27}$$

In equation 6-27, the photon moved from the rear of the instrument and toward the front. Its relative speed now is less. The total round trip time of the photon from the measuring point to the rear of the instrument and back to the measuring point is:

$$T_X = L_X / (C+V_E) + L_X / (C-V_E) \tag{6-28}$$

In Equation 6-28 we see that the round trip time of the photon depends upon the sum and difference of the light speed and the Earth's velocity.

Let us use V for the Earth speed in the X direction. As the photon enters the test instrument, we find the time for the round trip to be:

$$T_X = 2(L_X / C) / [1- (V/C)^2] \tag{6-29}$$

In equation (6-29) we find that the round trip time in the X direction has a correction term which involves the square of the Einsteinian correction factor. Thus:

$$T_X = 2(L_X / C) / \{[1-(V/C)^2]^{1/2}\}[1-(V/C)^2)]^{1/2} \tag{6-30}$$

In equation 6-30 the Einsteinian correction factor is shown twice for clarity. Let us now look at the Y direction. As worked out by Einstein, the distance the photon travels in the Y direction is longer by simple triangulation since it is moving along the hypotenuse of a triangle. The corresponding time is:

$$T_Y = 2 (L_Y / C) / [1- (V/C)^2]^{1/2} \tag{6-31}$$

Equation 6-30 and 6-31 are basically the Einsteinian equations. More complicated equations can be produced but since we are looking for a simple describing function for space and time, these equations suffice. Of course in order to see the truth of Einstein for orbital motion and the fallacy of Einstein for linear motion, his equations must be studied. For instrument balance:

$$T_X = T_Y \tag{6-32}$$

For this simple case in order for Time X to equal Time Y:

$$L_X / (A \cdot A) = L_Y / A \tag{6-33}$$

Where:

$$A = [1-(V/C)^2]^{1/2} \tag{6-34}$$

Thus:

$$L_X = A\, L_Y \tag{6-35}$$

Therefore:

$$L_X = L_Y\, [1-(V/C)^2]^{1/2} \tag{6-36}$$

In equation 6-36 we see that for all other factors being equal, the length (L_X) of the instrument in the direction of travel (X) is equal to Einstein's formula times the length (L_Y) in the perpendicular direction Y.

The Earth is physically shorter in the direction of travel by the Einsteinian correction factor. Since the Earth has a large mass, everything associated with the Earth will shrink in the direction of travel. If we take the instrument and move it 90 degrees, the X-axis will get larger and the Y-axis will get smaller. It will always maintain the size difference.

Einstein's relativity concerned a mathematical shortening of the X-axis. In true space-time there is an actual shortening of the X-axis and some possible enlargement of the Y-axis and Z-axis as well. True space-time is quite physical and subject to ordinary classical physics whereas Einsteinian space-time is purely mathematical and lacks some physical reality.

There is a time delay in these changes in the X and Y direction during rotation. If we rotate the instrument at extremely high speed and make it from a huge amount of mass, then the readings will change to show the effects of Earth speed verses the null point. Yet the original instrument is not designed to rotate at very high speed nor does it have a very large mass. Other correction factors will occur as well. However, things are quite different in pure free space. Upon the Earth we are locked into the mother gravitational field. Our instrument is only a tiny test instrument. It obeys the Earth's field. The Earth shrinks in the direction of motion and elongates somewhat in the perpendicular direction. We need a very massive instrument to beat Mother Earth's field. On the other hand once we move away from the Earth we will have good results.

In general the distances will change with velocity. The relative speeds of C+V and C-V will apply. Einstein chose one solution. We see that the results are the same for length for the Doppler space-time solution as for the Einsteinian solution, except that for the Doppler solution, the distances are a variable ratio of the arms rather than just shrinkage of the arm in the X direction.

In equation 6-36 we see that the moving object shrinks in the direction of motion as compared to the perpendicular direction. Yet, the perpendicular direction tends to elongate due to the motion. These are physical effects on large bodied objects. The atomic spacing between atoms within the instrument is subject to physical stresses due to motion.

The effects are non-linear but a steel pipe moving in the X direction will be compressed greatly due to the motion. The space-time equations show that not only will acceleration compress an object but also constant motion will produce permanent compression. If you look at the object as a ball of waves, if we move the ball in the X direction, the front waves will get distorted and compressed. The ball will shrink in the X direction and elongate somewhat in the Y direction. Since Einstein used constant Y and Z, this chapter will do the same so a comparison can be made. In the next chapter we will look at various solutions for the variation of Y and Z. Y can vary and Z can be constant, or both can vary, or both can be constant.

Einstein's relativity equations gave the wrong impressions. The moving object is the one that is compressed. The stationary object is not compressed. Our motion changes us with respect to the sun whereas the sun does not change.

Let us now look at the time clock and see what happens with a moving time clock. This will resolve the clock paradox as well. The clock paradox clearly shows the work of Einstein to be incorrect for linear motion. We can see this by looking at the mechanism by which the clock of a moving object slows down

Let us use the same M/M instrument as a clock. Let us keep the photon source within the instrument so that photons are constantly generated and move back and forth in the X-axis and Y-axis while being counted and measured. We then have a fancy instrument clock. The atomic clocks on a smaller scale will work similarly.

To keep the solution simple, let us consider the case where the speed of the instrument is rather low compared to the speed of light. Let us also assume that most of the distortion is in the X axis direction and that the Y axis distortion (bending) in the X direction is basically smaller than the distance traveled in the X direction. Otherwise we must use nonlinear analysis which would apply as soon as the velocity gets too large.

For the simple case, the instrument nulls and the time of travel in the X-axis equals the time of travel in the Y-axis.

$$T_X = T_Y \tag{6-37}$$

We can use the Y-axis as the reference axis. Thus:

$$T_Y = 2(L_Y/C)/[1-(V/C)^2]^{1/2} \tag{6-38}$$

Let $T_O = 2L_Y/C$, thus:

$$T_Y = T_O/[1-(V/C)^2]^{1/2} \tag{6-39}$$

We see that a moving clock mechanism takes more time to travel internally since the path length is longer due to motion. The clock slows as per Einstein.

Einstein was incorrect when he assumed that a clock on the ground would appear slower to a moving observer in the sky. If we look at a clock on the surface of the Earth and one in a satellite, we will notice that both are distorted and slowed due to galaxy rotation. Both are distorted and slowed due to the solar system motion. Both are distorted and slowed due to the rotation of the Earth.

The Earth is a large object. The rotating clock on the satellite does not disturb the gravitational field of the Earth. The distortions of the Earth clock and the satellite clock are common mode due to Earth rotation.

Now the clock in the sky has an additional velocity term due to its motion. In general the velocity vectors of Earth motion, sun motion, and galaxy motion are not simple vector additions due to the nonlinear distortions of all the dimensions. In general one must apply a root mean square solution to the velocities. The fact that the satellite is moving in the direction of galaxy rotation or against it does not readily show up since galaxy rotation is common mode to everything. However, more sophisticated rotating instruments with high angular velocity and heavy mass should show some differences.

The clock in the sky is slower than the clock on the ground since it has an additional motion above the common mode motions. They are both slowed common-mode due to all the other motions. If you took radar measurements of this including the Doppler corrections you would find this to be true.

If you were in the spaceship orbiting the Earth, the Earth's clock after Doppler corrections would appear faster. The clock paradox is not a paradox at all. It was just an error in thinking on the part of Einstein. Einstein applied steady state electrical theory to linear transient problems. It was a simple solution but it was only partially true.

Let us take two spaceships with radar system clocks and a ground station with the same clock. If the spaceships took off in opposite directions, they both would see each other's clock as moving the same after correcting for the Doppler effect. This will produce fewer and fewer pulses per second as the ships speed up. They both would see the Earth clock as moving faster after correcting for the Doppler. They could then travel quite a long time and later turn around fairly rapidly. Without considering accelerating and deceleration for a short turn around time compared to the total voyage time, they would start to pick up more clock pulses per second when they moved toward each other due to the Doppler. The time difference in the radar pulse counter would slowly head toward zero. They would still see each other's clock as the same and the Earth clock as moving faster.

Finally when they arrive once more upon the Earth, the two spaceships would find that the measured clock ticks of their own ship and the radar clock counter measuring the other ship are the same. The clocks would read exactly the same when they arrive upon the Earth. However the Earth clock would be faster.

Two ships traveling in opposite directions each see the other's clock as identical for identical speed. Of course in this analysis we assumed that the effect of the gravitational field of the Earth did not change the spaceship's clock. If we move far enough away from the Earth, we will get a change in the root mean square factors in the initial distortion of the clocks. There will be a tiny increase in clock speed due to being apart from the Earth. However the big distortion factor is the galaxy speed and this cannot be readily escaped. Yet, this is common mode to everything and all instruments tend to be distorted by it and it nulls out.

Let us now look at the mass of an object as it varies with velocity. In the original theory by Einstein, the relativistic mass of an object was computed using two balls that collided.

The first ball was on a stable reference frame such as the Earth and moved upward with a velocity V_{YU}. Forgetting about gravitational effects, a second ball was on a moving platform above the Earth. The platform was moving to the right with a velocity V. An identical ball was sent down from the moving platform at a downward velocity V_{YD}.

A perfect collision between the two balls occurred and the momentum in the Y direction was identical. The upward ball moved a distance upward of Y_O and the downward ball moved an equal distance. Since the momentum was conserved:

$$M_U V_{YU} = M_D V_{YD} \qquad (6\text{-}40)$$

In equation 6-40 Einstein reasoned that the upward momentum in the Y direction was equal to the downward momentum in the Y direction since both balls traveled an identical distance. In addition, both balls stopped at the halfway point and returned to their reference frames in an identical manner from when they started. The lower ball merely went up and down while the upper ball moved to the right while moving up and down. When we look at the lower ball we see that it moves a distance of only $2Y_O$ in a time T_O. The momentum of the lower ball is:

$$MV = M_U Y_O / T_O \qquad (6\text{-}41)$$

Things are a little different for the upper ball. It travels a much larger distance since the reference platform was moving to the right with a velocity V. Likewise it can be said that it travels the same distance in the Y axis direction but it takes longer to get there. The momentum then becomes:

$$MV = M_D Y_O / T = M_D Y_O / \{T / [1 - (V/C)^2]^{1/2}\} \qquad (6\text{-}42)$$

By equating the momentum of Equations 6-41 and 6-42, Einstein concluded that:

$$M_D = M_U / [1 - (V/C)^2]^{1/2} \qquad (6\text{-}43)$$

Equation 6-43 is Einstein's famous equation for the variation of mass with velocity. The rest mass is the upward mass and the moving mass is the downward mass.

The problem with Einstein's calculations was that it made no difference which was the reference frame and which was the moving frame. Each object according to Einstein would look at the opposite moving object as the relativistic mass. This is not correct. His universe was a strange mathematical non-physically realizable universe.

If you add energy to an object, its inertial mass increases. If you make it go fast it will have a large mass. If you look upon another object which you did not add energy to, it won't appear to have higher mass. Einstein's equations for special relativity are false for linear motion. Einstein's solution is correct for orbital motion where the reference is stationary such as the cyclotron or for planetary motion when the sun is used as the reference. As long as the reference is basically stationary, Einstein's equations are the best mathematical fit.

Einstein's equations are true when we consider the Earth as the reference platform. The Earth motions distort everything upon the Earth. Everything tends to be common mode. The moving clock above the Earth will slow. However our clock will look faster to it. The reason is that the Earth and the clock have been adjusted for galaxy motion, sun orbit, etc. The only thing different is that the clock above the Earth is moving with respect to us. That clock has already been adjusted to match the Earth. Thus Einstein's equations are correct as long as this Earth is the reference platform. The little clock only has a very tiny gravitational field. Therefore it is dependent upon the Earth's gravitational field for corrections.

It is also important to note that Einstein got the correct mass equation for the Y-axis with respect to the total mass on the X-axis. His orbital mass equation is true. However, he failed to recognize that moving a mass at the speed of light converts gravitational mass in the X direction into inertial mass.

SECTION 6-3: DOPPLER MASS CENTER OF GRAVITY

Let us now look at the concept of the Doppler mass center of gravity. Let us investigate a huge ball that is perfectly spherical moving in an area of free space between galaxies. If the ball is perfectly stationary in the universe, it will be perfectly round.

A three-dimensional Doppler laser gyroscope within the ball will always null and it will have an offset error if moved in any direction at any velocity. When initially perfectly stationary, the device can be calibrated by moving it in each of three planar rotations to see that it is equally balanced. Later when moved it can be calibrated for linear motion. The accuracy of the instrument depends upon the magnitude of its own gravitational field as compared to the gravitational field of the surroundings. An ideal instrument would be heavy and would be between galaxies. It should work well with a 10/1 ratio of self-gravitational field to that of the surroundings. An ideal ratio would be 10,000 to 1 for near perfect results.

The gravitational field of the perfect ball will extend outward for millions of miles producing a huge electromagnetic ball. It will be perfectly spherical. The ball and the gravitational field will both be perfect. However, the field is a spectrum of individual atomic fields. It is a perfect but highly complex sum total of billions of individual fields operating at a spectrum of frequencies.

Let us now push the ball and let go. Initially the ball will compress at the point we pushed it. This will move all over the ball and to the gravitational field itself. Later when we achieve stability, we will find that the ball is permanently distorted in the direction that it is pushed. It will also show some distortion in the perpendicular direction as the pressures equalize. It gets shorter in the direction of motion but longer perpendicular to the direction of motion. The gravitational field will change as well. It will react outward from the push at the speed of light.

The inertia of the object is caused by the displacement of all the current loops in the present from the past and the force generated by Ampere's law. In addition, the atomic spacing will be effected causing shrinkage in the direction of travel.

The pushing force will compress the ball in the rear as a transient condition. Later in steady state the front will compress and the rear will spring back toward normal. A permanent shrinkage will occur.

Eventually, the internal gravitational field will have compressed waves in the front of the motion and elongated waves in the rear of the motion. The ball will then have a Doppler mass distribution. The density of the mass will be greater in the front and less in the rear. The Doppler mass equations are:

$$\text{Mass front} = M_X \, C/(C-V) \tag{6-44}$$

$$\text{Mass rear} = M_X \, C/(C+V) \tag{6-45}$$

The Doppler geometric-mean mass, which is the inertial mass, calculates to be:

$$M_{Xi} = M_X / [1 - (V/C)^2]^{1/2} \tag{6-46}$$

Let us look at the case where the ball is moving rather slowly compared to the speed of light. We can then approximate the front and rear Doppler masses as follows:

$$M_F = M_X [1 + (V/C)] \text{ approximately} \qquad (6\text{-}47)$$

$$M_R = M_X [1 - (V/C)] \text{ approximately} \qquad (6\text{-}48)$$

If we look at the front mass at a distance of $(+L/2)$ from the center of the ball and the rear mass at a distance of $(-L/2)$ from the center, the new center of mass is:

$$\text{Center of Mass} = +L\,(V/2C) \qquad (6\text{-}49)$$

In equation 6-49 we find that a body moving at modest speed has a shifted center of mass equal to half the length of the body times the ratio of the velocity of the moving body to the speed of light. This is for a body such as our Earth that moves slowly compared to the speed of light.

The faster the body moves, the closer the center of gravity will move toward the front surface. For high-speed objects, we would get another Einsteinian type equation with Doppler corrections for the center of gravity of the object. Yet for our Earth and most objects, equation 6-49 will suffice.

If we look at our Earth we will find that the axis of the Earth will be off centered from the true center of gravity. This in turn is offset from the physical center due to an uneven distribution of mass. The axis of rotation of the Earth is off centered from the stationary center of gravity due to the motion of the Earth. There is a Doppler shift as per Equation 6-49 in the center of gravity of the Earth.

As the Earth orbits the sun, its velocity increases and later decreases. The center of gravity will shift every day. This will cause the Earth to tilt. We get summer and winter due to the Doppler shift of the center of gravity of the Earth.

The change of the center of gravity of Mercury should be more severe as it moves closer to the sun. Einstein calculated the motion of Mercury from his equations whereas no one else could. His space-time equations have always been a good fit. However, Einstein never brought to mind the Doppler mass effect. Although his equations are good, the real reason for the motion of this Earth and Mercury as well is that the center of gravity changes as the velocity of the planet changes. In addition, changes in position toward or away from the sun result in perpendicular forces as the gravitational fields interact with each other.

Our moon is phase locked to us like a synchronous motor due to the offset center of gravity. The galaxy motion causes center of gravity distortions in our sun, our Earth, and our moon. However, this may not be easy to detect since the radius around which the distortions are occurring is so far from us. Since both the Earth and sun are moving within the galaxy, it may not be so evident that the Earth's axis has an offset component due to galaxy motion.

Our moon will have an offset axis due to our motion around the sun and its motion around us. We appear phase locked to the face of the moon. If the moon was perfectly uniform and there was no Doppler shift, the moon should rotate. However the moon is not quite uniform. It is subject to a Doppler shifts, due to rotation around us and around the sun. The forces acting upon the moon tend to produce a phase locked condition after awhile.

The Doppler shift in the center of gravity of the Earth also tends to add heat energy to the core. The energy of rotation is slowly turned into additional heat in the core of the Earth. The Earth very slowly moves toward the sun over a long period of time. The rotation of the Earth also tends to slow. Eventually in the far future one face of the Earth will end up facing the sun. Then the Earth will no longer support life, as we know it.

SECTION 6-4: DOPPLER RELATIVITY REFERENCE SYSTEM

Let us look at an object such as a perfect sphere that is rotated upon an axis until it reaches near the speed of light at its circumference.

Einsteinian relativity found that a man in an elevator moving at a constant velocity V could not detect his motion. Yet, if the man were spinning with a constant rotational velocity, he would be pinned to the outer surface of a round or spherical elevator. Rotational motion can be detected but not linear motion as per Einsteinian relativity. This is a fallacy.

True Doppler relativity does not distinguish between linear motion and rotational motion. Both can be detected. Rotational motion is easier to detect and linear motion is harder to detect. A three-dimensional Doppler gyroscope set in pure free space will measure velocities elsewhere. It will not null upon this Earth, however once in our gravitational field, it will become distorted.

Perfect instruments must be kept away from strong gravitational fields; otherwise they will exhibit errors. It is difficult to overcome the distortions of the Earth itself. Once we null the instrument here on Earth, it will tend to always null. However, if we null it in outer space then it will not null here.

In a universe where all the planets and stars are moving and the entire universe may be rotating as well, a reference point is hard to find. Of course for a possible multi-light-speed set of coexisting universes, we can use the highest light speed universes as the reference. We can imagine our universe defined in terms of distances and time upon the highest light speed references. However for ourselves we need to know what the reference is for our own universe.

The basic reference for the universe is speed. If we take an object in pure space far from any galaxy or between galaxies, the forces upon the object will be nearly zero. Let us take a sphere and place it within a larger sphere that reflects the light back to the inner sphere. Since we are far from any galaxies, we exist in perfect space. Let us beam light in all directions from the inner sphere to the outer sphere and obtain a null by firing little rockets on the outside of the sphere until we get a perfect null pattern. When the Doppler signals in all three axis, return with a perfect null, then the spherical spaceship is perfectly stationary.

As we move the spaceship with a velocity V with respect to a ruler on the spaceship, we will find that the null is destroyed. If we try this upon the Earth, we get problems from gravitational distortions due to the motion of the Earth, since we are the Earth and not the instrument. The perfect instrument will have corrected lengths that tend to negate easy measurements. In perfect space we do not have problems with nearby gravitational fields. For an instrument to work properly, the gravitational field of the instrument itself must be much larger than the external gravitational fields of the surroundings. A ratio of ten thousand to one would be ideal. However instruments should work with ratios as little as ten to one.

The M/M experiment exists within the Earth's gravitational field and lacks the ability to test for zero speed. The mass of the test instruments must be free of external influences. This is clearly an impossible problem for experiment upon this Earth. The best we can achieve is experiments on a space ship at a gravitational null point between our sun and another local star. Everything else is too far to readily accomplish.

Our spaceship can be nulled to obtain perfect rest. Then we can measure the Doppler null shifts to determine our speed. As long as we are moving slowly compared to the speed of light, we remain fairly linear.

As we speed up, our spherical spaceship will flatten in the direction of motion and expand perpendicular to the direction of motion. We will become distorted. The solution to the Michelson/Morley experiment shows that the ratio of the two perpendicular axis changes as the object moves in orbit.

The distortions and the Doppler null shifts tell us our speed. Thus Doppler relativity shows distortions as speed increases. It also shows that speed itself is the reference in our universe.

If you look at Einstein's equations you will find that they all include the velocity V in them. Distance and time change but V remains the same. Thus, the true reference system is the velocity V.

The primary principles of Doppler relativity state that:

DOPPLER RELATIVITY PRINCIPLE 1:

ALL NON-ROTATING LINEAR REFERENCE SYSTEMS MOVING IN PURE FREE SPACE WITH THE SAME ABSOLUTE VELOCITY (V) ARE EQUIVALENT.

Principle number one states that all linear non-rotating reference systems moving with the same absolute velocity are equivalent. The rulers, the distortions, and the time clocks of these reference systems will be identical. If the two systems are moving in the same vector direction, radar signals between the two will show perfect clock frequencies. If the two systems are moving in opposite directions, the Doppler clock measurements will be lower identically for both systems and spaceships. If the two spaceships are moving toward each other, then the clock frequencies measured will both be higher.

If you have one space ship with a particular absolute velocity and another with a different absolute velocity, the two systems are not equivalent. Both will have different rulers and clocks. If the velocities are reasonably slow relative to the speed of light, then we can calculate the various rulers and clock frequencies and obtain the Doppler radar calculations. These will require more complex calculations than for equivalent systems.

If one space ship is moving at half the speed of light, things get very complicated, as the distortions are very large. The error that Einstein made in relativity is that he assumed that the speed of light was the only reference. The speed of light is one reference and the absolute speed of an object is the second reference.

Let us now look at a rotational object. We see that all reference systems that move at the same absolute velocity are equivalent. This is difficult to produce except between galaxies. However, if we look at two galaxies which are moving at the same rotational speed and look at two planet Earth's which are in the same location, then we can say that the two planet Earth's will produce identical results. In general except in free space it is difficult to find two objects with identical speeds.

Of course when all objects are moving slow compared to the speed of light, we get good results by assuming that one object is stationary. Let us now take a solid ball far out in space and start to rotate it. Far in space, all objects rotating with the same circular momentum are identical if their linear velocities are the same. Thus:

DOPPLER RELATIVITY PRINCIPLE 2:

ALL REFERENCE SYSTEMS MOVING IN PURE FREE SPACE WITH THE SAME LINEAR VELOCITY AND THE SAME ROTATIONAL VELOCITY ARE IDENTICAL.

Rule No 1 was for non-rotational linear systems whereas rule No.2 includes rotations. Of course if we are on a fast moving rotating spaceship we will know it pretty rapidly. The human body can take steady linear motion up to nearly fifty percent of light speed without problem. However, once we rotate too fast, we will just perish.

Let us rotate the perfect sphere. As we rotate the sphere, it will get longer perpendicular to the axis of rotation. It will also start to shrink along the axis. This is just simple strength of materials mechanical theory. We get distortions similar to the Michelson/Morley experiment within the ball.

As we add more and more rotational photon energy to the ball, it will get longer perpendicular to the axis of rotation and it will flatten at the poles. The mass of the ball will continue to rise. A clock within the ball will slow as per an equivalent Einsteinian or root mean square Doppler time formula. The length along the axis will shrink as per a similar Einsteinian formula.

Although the length along the axis shrinks, the diameter expands greatly. During the rotation of a perfect sphere we find that both the axial and transverse dimensions change.

As the speed of the rotating sphere approaches nearly the speed of light on its circumference, its diameter expands greatly and it's inertial mass approaches a large number. In addition, its length approaches zero.

We see how the rings of Saturn were formed and how galaxies can move from flat planes to more spherical shapes. If you take energy moving close to the speed of light in a circular plane, it will look like a flat disk spinning rapidly. As the disk slows, it will start to form a spherical shape in the middle. As we move further away from the center, the velocity will approach the speed of light. As we move close to the center the velocity will approach zero.

If we spin the Earth faster and faster, it would flatten at the poles and elongate perpendicular to the axis of rotation. We would then produce rings around our Earth. As the Earth slowed, the rings would either turn into a moon or several moons or remain as rings for awhile.

Very high rotational forces formed Saturn. The same was probably true of our solar system. Since it operates in one plane, it probably was formed from a large ring around the sun moving close to the speed of light. Yet, many alternate possibilities exist. The only thing important is that a large spinning disk can turn into a sun and planets. Photon energy spinning in a plane can become a solar system or a galaxy for that matter.

Einstein's relativity turns everything into a simple problem. All reference frames are equal. Yet, that is not true. The only times things are equal is when we have identical rotational velocities and identical axial and transverse velocities. Of course we must then be free of any major stationary gravitational fields or moving gravitational fields.

The only conclusion is that all reference planes in the entire universe are unequal. Here and there you will find reference planes which are almost equal to each other. Even upon the Earth, you will not find two reference planes that are absolutely equal. The closest you can get is two reference frames in the same laboratory Thus we get:

DOPPLER RELATIVITY PRINCIPLE 3:

NO TWO REFERENCE PLANES IN THE ENTIRE UNIVERSE ARE ABSOLUTELY IDENTICAL

Unfortunately we cannot find two reference planes in the entire universe which have the identical conditions. Even if you could find two spaceships with the exact rotation and linear velocity, they will still exist with gravitational fields near them that are moving and different.

Einstein though that all reference-systems were identical and could be tied to the speed of light. In truth, we live in a universe where you can find nothing absolutely identical. The best we can get is areas of pure free space far from stars and the centers of galaxies. On the other hand, this planet Earth moves relatively slowly so, it is a great reference itself.

SECTION 6-5: LENGTH & TIME INVARIANCE

In GG/MM/Doppler space-time, mass is invariant. Although there are changes from gravitational mass in the X direction to inertial mass, the absolute mass remains the same. Let us now look at length and time in GG/MM/Doppler space-time.

For the simple linear solution, L_Y and L_Z remain constant. For this solution, the only thing that changes is the length in the X direction or L_X. Let us look at the total Doppler length in the X direction. To the right and left it is:

$$L_{XR} = [L_O [1-(V/C)^2]^{1/2}] \cdot [C/(C-V)] \qquad (6\text{-}50)$$

$$L_{XL} = [L_O [1-(V/C)^2]^{1/2}] \cdot [C/(C+V)] \qquad (6\text{-}51)$$

The RMS value of the Doppler is:

$$L_{XRMS} = L_O \qquad (6\text{-}52)$$

As the velocity reaches light speed, we see that the Doppler length to the right moves toward plus infinity and the Doppler length to the left reaches zero speed, while the RMS Doppler is steady at L_O. The image of an object stretches out toward infinity in the forward direction and zero rearward but the object remains the same size from a root mean square describing function analysis.

The importance of Equation 6-50 in spacecraft design is that the fast forward image or Doppler right image of a strong pointed nose cone reaches outward toward infinity. This will destroy or deflect any light object in front of the spacecraft. You do not have to worry about a pebble hitting the spacecraft and damaging it. The proper design of the spacecraft is its own safety feature.

We see that length is invariant in true space-time. Both mass and length are invariant and quite classical for linear space-time. Now let us look at the time clock. The time to the right and left are:

$$T_{XR} = [T_O [1-(V/C)^2]^{1/2}] \cdot [C/(C-V)] \tag{6-53}$$

$$T_{XL} = [T_O [1-(V/C)^2]^{1/2}] \cdot [C/(C+V)] \tag{6-54}$$

The RMS Doppler time is:

$$T_{XRMS} = T_O \tag{6-55}$$

We see in equation 6-55 that the RMS Doppler time is also invariant. Therefore space, time, and mass are all invariant in true linear space-time for the constant Y/constant Z solution. Einstein's solution for Orbital space-time is not applicable to linear space-time. Only classical techniques are proper. In Chapter 7, other solutions will be investigated with variable Y and variable Z.

SECTION 6-6: CHARGE INVARIANCE

Let us look at the invariance of charge with velocity. The electrical charge Q of the electron will change when it moves. It will experience a Doppler effect. This will give rise to the magnetic field of the electron. Just like mass and length the Doppler equations for charge are:

$$Q_{XR} = [Q_O [1-(V/C)^2]^{1/2}] \cdot [C/(C-V)] \tag{6-56}$$

$$Q_{XL} = [Q_O [1-(V/C)^2]^{1/2}] \cdot [C/(C+V)] \tag{6-57}$$

$$Q_{XRMS} = Q_O \tag{6-58}$$

We see that the charge Q behaves the same as the mass. It is invariant for the constant Y/constant Z solution. There will be a Doppler charge ahead of the root mean square or gravitational charge. The charge Q of the electron will exist in front of itself, and the moving charge produces the magnetic field of the electron. When an electron is spun in a circle, we have an Einsteinian orbital effect. The charge for orbital motion is:

$$Q = Q_O / [1-(V/C)^2]^{1/2} \tag{6-59}$$

We see in equation 6-59, the build up of charge for orbital motion. The faster the electron moves, the greater the charge builds up. Of course we define this effect by the magnetic equations. The magnetic field is the result of the Doppler effect on moving electrons. In addition, it is the charge buildup due to repetitive cycles of the electrons.

For linear motion, most of the magnetic effect is spread in front of the electron and behind the electron. The perpendicular charge of the electron in pure free outer space tends to remain fairly constant just as the mass does. There is always some variation however.

When we look at a transmission line we find a strong magnetic field surrounds it. Of course this is still an orbital situation since the current flows from one wire to a load and back through the other wire to the generator. Thus it is not a free space problem.

In later chapters, the Doppler Charge of the dot itself will be explained. For now, once we move a charged particle, we get a Doppler magnetic field. The magnetic field is merely the Doppler electric field.

CHAPTER 7 GG/MM/DOPPLER ORBITAL SPACE TIME

SECTION 7-0 INTRODUCTION

In Chapter 6, Doppler linear space-time was added to Michelson/Morley space-time to produce GG/MM/Doppler space-time for linear motion. The equations of Einstein appear quite reasonable for problems dealing with orbital space-time. In Chapter 6, we constrained the Y and Z-axis to be invariant. This enabled us to compare the GG/MM/Doppler solution with the Einsteinian solution. We could then see where Einstein got into trouble with his clock paradox problem. As seen in the last section, true time is invariant and no clock problem exists.

The man in the spaceship is slowing down but the true real time, which includes the RMS Doppler, does not change anywhere in the liner portion of the universe. In the linear region of the universe, the true clock, true ruler, and true mass are invariant. Of course a man in a space ship moving at 1000 miles an hour somewhere in the universe gets different clock reading than a man moving at one million miles an hour. However, the total time including the Doppler is the same for both cases for the Constant Y/Constant Z solution.

There certainly are areas of space where the light speed reaches toward zero and things will never be the same as here. However, all areas of reasonable linear space and all planets such as this Earth will read approximately the same things.

We were forced to rely upon the Michelson/Morley experiment since it is the only thing we have. Since we are tied to mother Earth and are in orbit around the sun, the test instrument is denied true linear space-time readings. Therefore we get readings applicable to both orbital motion and true linear space-time as well. This leaves an area of uncertainty in the experiment. It also enables the scientists who read the data to be fooled. It is like being flat creatures in a three dimensional world.

Classical physics has worked for hundreds of years. Therefore true classical space-time including the Doppler effect can provide the same answers as orbital space-time. In this Chapter we will look at the equations for orbital space-time from a classical perspective. In particular, the only real differences between linear space-time and orbital space-time are in the mass/energy equations.

From a classical viewpoint we will explain why the mass/energy rises toward infinity as the velocity approaches the speed of light in the cyclotron. We will also explain why Einstein's equations work well for the Bohr Orbit and planetary motion.

The slowing of the clock is common to both GG/MM/Doppler linear and orbital space times. Time is invariant once we add the Doppler RMS value. The time clock is the same all over the universe. The same is true of the ruler. The ruler shrinks but the Doppler rms length is the same. For linear space time, Mass is invariant while for Einsteinian Orbital space-time mass rises toward infinity as we move upward in light speed. Now we need to look at the rise in mass with velocity for orbital motion from a classical perspective.

SECTION 7-1: GG/MM/DOPPLER ORBITAL MASS EQUATIONS

Let us understand what happens when an object is moved in pure free space. We push an object in the X direction. The object shrinks in atomic spacing in the X direction at the point of pushing. This translates throughout the object. Each proton resists the pushing. There is a force between the proton inner orbit of the present cycle and the memory of the past cycle. Ampere's laws of current loops apply. Billions of spherical orbits of the present are held back by billions of spherical orbits of the past complete cycle.

When you push the object, the push moves through the object at the speed of light toward the front protons, neutrons, and electrons. Eventually we get a distribution of forces from the rear to the front. There will be shrinkage of the object in the X direction. When the force is removed, the Doppler effect will cause the front of the object to operate at a higher frequency than the rear. The object will be crushed in the front more than the rear. The shrinkage will not be as great in the steady state as during the initial transient pushing. However, it will be the front that is more crushed than the rear.

As we look at each proton, we will see a displacement between the present complete orbit and the past complete orbit. This will be a permanent displacement. The same will be true of the neutrons and electrons. The faster that we move toward the speed of light, the greater will be the time and distance shift. Thus the proton and its image will not match for a moving object.

As we look at the inner oscillations of the protons, electron, and neutrons, and the Bohr orbit itself, we find that perfectly spherical oscillating orbits in 360 degrees by 360 degrees patterns will start to flatten. The orbits will start to turn into circular orbits perpendicular to the motion in the X direction. We start to turn the X dimension into photons. The gravitational mass drops toward zero in the X direction but remains approximately the same in the Y and Z directions.

In chapter 6 the Michelson Morley equations showed no increase of mass in the Y and Z directions for an object traveling in the X direction. However, we see that the turning of all the orbits perpendicular to the direction of motion will increase the mass/energy somewhat. This is accounted for by the increase of mass/energy due to added photon energy. From classical physics there will be a degree of increase of mass with velocity. Let us look at the energy formula:

$$E = M_0 C^2 + \tfrac{1}{2} M_0 V^2 + A M_0 C^2 \tag{7-1}$$

In equation 7-1 we see that the increase of mass/energy of an object moving in a linear path is equal to its initial energy plus its kinetic energy plus its increase of mass with velocity. We see that the turning of the individual orbits from a perfectly spherical shape to a flat shape decreases the gravitational mass in the X direction to zero and slows the orbit clocks.

The combination of the slowing of the orbit clocks and the flattening of the orbits tend to compensate for each other. However, as we move closer and closer to the speed of light, we start to get non-linear effects. Equation 7-1 provides us with a describing function, which includes all the non-linear effects. The net result is that in order to achieve 100% light speed we get:

$$(3/2) M_0 C^2 < E < 2 M_0 C^2 \tag{7-2}$$

In equation 7-2 we have (A) less than or equal to ½ which produces energy of between 3/2 to 2 times the rest mass to bring an object up to exactly light speed. Since the mass in the X direction drops with velocity, we get fancy integrals. However, it is difficult to take into account the Fourier series and harmonics involved. Equation 7-2 is typical for electrical theory. There is always an additional potential energy term in everything. When we achieve light speed we have a potential energy to overcome which equals at a maximum the kinetic energy. This potential energy overcomes the final harmonics and Fourier series distortions. We can then get pretty close to light speed with 50 percent fuel. However we need up to 100% fuel to get to light speed where the fuel equals the mass of the spacecraft.

In Einsteinian space-time, the speed of light is held constant. According to Chapter 3, the speed of light for the universe has an extremely small but very important variation over the cycle time of the universe. For all practical purposes, the speed of light is constant. However, that is the ideal speed of a photon traveling in pure free space away from any interference or gravitational fields.

The speed of photons reaches close to zero in black holes. It is small in diamond. It varies with the moving gravitational field. The photon changes from pure inertial mass in free space to almost pure gravitational mass within the electron or proton. The speed of light as measured by photon energy is a variable.

One big problem with relativity is that the speed of two photons moving apart is kept to the speed of light C. This is a false concept. Two photons moving apart from each other move apart at a rate of 2C. Lorentz and Einstein looked at the universe as an entity, which obeyed general electrical theory laws. However they mistook the relationship between an object and the universe and two different objects. Each object is limited to a speed of C with respect to the universe but each object can have a speed of up to 2C with respect to other objects.

They also made everything in the universe its own reference frame. They failed to recognize that a zero velocity Doppler gyroscope can be built and absolute speed can be determined. Right now we must rely upon red shift data from the astronomers and background radiation since big bang. Of course if you can find identical radio stars then you can use their Doppler frequencies to determine our speed as well. The problem is that the Earth's massive gravitational field limits us. The velocity of an object can be:

$$0 < V_{B1} < C \qquad (7\text{-}3)$$

In equation 7-3 we have a ball (B1) which has a velocity greater than 0 and less than the speed of light C. The same is true of a second object or ball (B2). Thus:

$$0 < V_{B2} < C \qquad (7\text{-}4)$$

Each ball or object is constrained by the universe to move at less than or equal to the speed of light C. Of course the speed of light depends where you are. In pure free space it will be Co or the ideal light speed. Within a black hole it will be zero.

The relative speed between ball (B1) and ball (B2) is:

$$-2C < V_{B1} - V_{B2} < 2C \qquad (7\text{-}5)$$

In equation 7-5 we see that two balls can be moving apart at twice the speed of light even though each ball is constrained by the universe only to travel at the speed of light C.

Einstein's theory forces the balls only to have a maximum relative speed of C. This produced a very constrained set of equations which were reasonably good for orbital motion but which produced the clock paradox and turned the minds of man upside down.

On the positive side for an Einsteinian viewpoint, when you look at orbital motion for a steady state Doppler solution, mass/energy is built up within the repetitive loop. The Doppler to the front reaches the Doppler to the rear and visa versa. The net result is:

$$M_{XF} = M_O C / (C-V) \qquad (7\text{-}6)$$

$$M_{XR} = M_O C / (C+V) \qquad (7\text{-}7)$$

In Orbital space-time, the Doppler in the front fills the entire orbit with waves. This returns to the rear of the mass. It then goes forward in a steady state build up. For each cycle that the mass moves, the Doppler is moving at the speed of light around the loop. The net result that the gravitational mass of the object in the X direction does not drop as it did for linear space-time motion. The entire orbit becomes the mass. This produces Einstein's solution.

Within the cyclotron, the electron exists all over. We have a line of charge filled with a huge amount of energy. The same is true for the Earth's orbit. The Earth in orbit for billions of years has a Doppler space-time image. The orbit time for the image is:

$$\text{Image Time} = \text{Circumference} / \text{Speed of light} \qquad (7\text{-}8)$$

For rough numbers, the sun is 93 million miles and the speed of light is 186,000 miles per second. Thus the image time of the Earth is:

$$\text{Image Time} = 2\pi\, 9.3E7 / 1.86E5 \qquad (7\text{-}9)$$

$$\text{Image Time} = 1000\pi \text{ seconds} \qquad (7\text{-}10)$$

It takes 3.14159E3 seconds or 1000 π seconds for the Doppler image to circle the sun for the first time. The rough numbers are rather interesting from a numerical point of view. In any event within about 3000 seconds, the first Doppler wave traveled the entire orbit. After it repeats four times, the entire image of the Earth has reached a steady state condition. It doesn't take a billion years to do this. The image stability should follow an e^x curve. Four time constants should do it.

$$M = [M_O / [1-(V/C)^2]^{1/2}] \cdot [1 - e^{-(t/T)}] \qquad (7\text{-}11)$$

where:

$$T = 2\pi R_S / C \qquad (7\text{-}12)$$

In Equation 7-12, we see that the image of the Earth has a time constant of the Earth's circumference over the speed of light. In Equation 7-11 we see that the transient orbital image mass is the Einsteinian mass times the standard electrical theory ($1 - e^{-(t/T)}$) function. It takes approximately 4 time constants for stability. If the Earth were traveling at the speed of light the time for stability would be:

$$T = 2\pi R_S / [C^2 - V^2]^{1/2} \qquad (7\text{-}13)$$

We see in Equation 7-13, that the time for image stability of a planet moving at close to the speed of light is quite large. It approaches infinity. This equation is also applicable to the Bohr orbit. It takes time to stabilize the Orbits. It takes time to move from a neutron state to a hydrogen atom. By using the above equations we can calculate many different time constants. The Earth also has a long-term physical cosmic time constant as it slows in orbit. This is very much larger than the image time constant.

Einstein's Orbital equations work well because of the build up of standing waves of mass/energy within the orbit. The image of our Earth exists all over our orbit. Therefore our Earth is bigger than we think. It is the entire orbit.

In Chapter 8, we will use similar equations to show how the orbital clock slows with velocity. This is merely RMS Doppler with a steady state solution. Energy builds up and clocks slow in the steady state. Stability comes fast, by the time an airplane reaches the sky the light waves have already traveled around the Earth a huge amount of time. The airplane produces standing waves and moving standing waves in the sky. The image of the airplane then exists all over the Earth. Its mass increases as per Einstein due to the Doppler standing waves. Which produce the image effect.

If you took the same airplane with a rocket engine in pure outer space it would work just like any classical physics mass until you get very close to light speed. The real difference between classical physics and Einsteinian space-time is the difference between electrical transient analysis and orbital steady state analysis.

An object in pure free space does not build up a Doppler image which returns to it. There is no steady state mass build up or standing waves of energy to fight. When you push the electron in the cyclotron you must push all the moving standing waves of energy in front of it. The more you push them, the more you produce them. Thus you build up a huge amount of energy in front of the electron which joins with the electron. In the process, you produce a Doppler image of the electron.

The electron never really gained mass/energy. All that happened was that a huge photon field of energy in a circle was produced. You can build this up to infinity. In pure outer-space you don't have this problem. All you have is the little spacecraft with its relativity tiny mass. It develops a Doppler mass in front of it and behind it but this is like a flashlight beam. It exists but it does not feed back upon itself. Therefore there is no mass build up to speak of except the small regular classical physics energy necessary to move it. You do not produce an image of the spacecraft in pure free space.

The big problem with regular physics is that we live in a purely electrical universe. You have to be an electro-physicist to really understand space and time. Einstein was fooled by steady state electrical theory verses transient electrical theory. He got great answers but failed to understand what was happening. Electrical theory permits a lot of interesting phenomenon to occur. We can get images in space-time. We can get inversions in space-time. We can produce Thevenin equivalents of space-time.

Let us return to the Michelson/Morley equations. The ratio of the length of the X-axis of the M/M apparatus to the Y-axis of the apparatus when the instrument is moving in the X direction is:

$$L_X = L_Y [1 -(V/C)^2]^{1/2} \qquad (7-14)$$

$$L_Z = L_Y \qquad (7-15)$$

In equation 7-14 we see that the solution to the Michelson/Morley experiment required that the X-axis shrink during motion in the X direction and that the Y-axis remains constant, enlarges a little, or shrinks a little. This was one of the alternatives of the M/M experiment discussed by various scientists long ago. It is a most simple solution and a real physical solution as well.

Let us take a ball of radius L_X that is perfectly stationary in pure free space, and compare it to a light sphere pulse emitted from it.

The absolute velocity of the ball is V. Under these initial conditions:

$$L_X = L_Y = L_Z \qquad \text{(for V=0)} \qquad (7\text{-}16)$$

In Equation 7-16 we see that the radius of the sphere is the same in all directions when the sphere is not moving. If we have the sphere move with the absolute velocity V with respect to the center of the light sphere, we find:

$$L_X/L_Y = L_X/L_Z = [1 - (V/C)^2]^{1/2} \qquad (7\text{-}17)$$

$$L_Y = L_Z = L_O \qquad (7\text{-}18)$$

In Equations 7-17 & 7-18 we see that as an object moves in the X direction, the X radius shrinks and the Y and Z radius remain constant. If we continue moving faster and faster, as we reach the speed of light, we will hit non-linearities and the Y and Z dimensions of the ball will get either somewhat larger or shrink while the X dimension moves toward zero. This occurs when we have a high density of dots such as the proton itself.

The exact point at light-speed requires a more detailed study and analysis. It may be possible to shrink the Y and Z dimensions, which would insure survival of a future light-speed spacecraft. If the Y and Z dimensions enlarge at light-speed, this could be a severe problem. However, if it could be made to shrink using a strong electromagnetic force field, than the light speed barrier could be achieved or broken safely. It might even be possible to ride slightly above the present light speed in the light-speed hysteresis loop. This work is for the future.

If the mass was a photon, it would only have a low density of dots and would only expand slightly or shrink in the Y and Z direction as it achieved light speed. The X direction will be at the minimum size, which is pretty close to zero. Thus when a spherical ball of energy within an electron becomes a photon, it becomes a plane wave of zero thickness in the X direction and limited size in the Y and Z dimensions.

GG/MM/Doppler relativity states that objects change their shape when you bring them upward to the speed of light. First they will flatten in the X direction. Later they will elongate slightly in the Y and Z directions. Alternatively they could be forced to shrink in the Y and Z directions. When the speed of light is reached, they will be turned into a flat circular plane of zero thickness. Thus a gravitational sphere of pure dot energy in all directions becomes a non-gravitational plane of light in the X direction and a gravitational mass in the Y and Z directions.

SECTION 7-2: INERTIA IN M/M SPACE TIME

Let us look at the property of inertia in GG/MM/Doppler space-time. In addition, we will look at centripetal and centrifugal forces in space-time. In general, the laws of inertia have been written down but no explanations have come with them. Einstein's space-time has been a set of mathematical rules without any explanations as well.

In GG/MM/Doppler space-time, the explanation becomes self-evident. Since GG/MM/Doppler space-time is really classical physics extended, everything learned in classical physics becomes part of space-time. In Einsteinian space-time, one had to divorce oneself from one's prior learning in classical physics. Classical physics always had the advantage of being understandable to the average Engineer and Scientific reader. Einsteinian space-time was not readily understandable.

Let us look at a cube of material of 1 meter in all dimensions. Let us place this material in pure outer-space and at pure rest. The absolute dimensions on all sides are one meter is all directions. The initial velocity is zero in all directions.

Let us now push on the cube to the right with a force F. We know that:

$$F = M A \qquad (7\text{-}19)$$

In Equations 7-19 from classical physics we see that if we apply a force (F) to a mass (M), it will accelerate with acceleration (A). Yet, the moment we apply the force, it does not accelerate. Equation 7-19 is a steady state type equation. It is a mathematical equation that covers up a more complex operation. Thus the basic equation of physics is a mathematical equation derived from mechanical scientists and mathematicians but not Electrical Engineers or Electro-physicists. Equation 7-19 tends to cover up the truth. It does not show what is happening in reality.

As we look inside the block of material, we see electrons, protons, neutrons, and photons, all moving about in various motions. Each has a space-time pattern that extends far beyond the block. The gravitational wave-patterns are quite complex. It is a spectrum of billions of spherical AC fields of all the electrons, protons, neutrons, etc.

We see that the block's gravitational field spectrum extends to huge distances. The extended block is the same size as the entire universe. When we push on the block we will be exerting a force on the entire universe. When we push on the block on the left side toward the right, the left side will start to compress.

The initial reaction of the block was to compress at the left side of the object while the right side remains unaffected. This is at a time prior to the motion of the object. This is shown below.

F>[....................*................................]   Figure 7-1

In Figure 7-1, we have a force (F) pressing on the left side of the object and compressing the object very much on the left side and not at all on the right side initially. The Center of Gravity noted by the star (*) will be on the left of the center. The center of gravity of the mass will be shifted from the exact center to the left of center due to the physical compression of the mass. This becomes a physical Doppler Mass center of gravity shift. Each proton, neutron, and electron resists the motion and the force moves through the object.

The object starts to move. As it develops velocity, we get a space-time center of gravity shift due to the compression of the gravitational waves to the right and the elongation of the waves to the left. Although each atom is involved throughout the length, the net effect is that the object will appear compressed in the front in the steady state and elongated in the rear. For small velocities compared to the speed of light, the approximate Doppler formula can be used. Thus:

$$\text{Doppler Mass Right} = M_O (1 + V/C) \qquad (7\text{-}20)$$

$$\text{Doppler Mass Left} = M_O [1 - (V/C)] \qquad (7\text{-}21)$$

The center of gravity due to the space-time Doppler moves:

$$\Delta L = L_0 (V/2C) \quad (7\text{-}22)$$

The simple linear approximation equation 7-22 shows that the center of gravity due to the Doppler space-time moves from the center toward the right edge as the object increases velocity.

We see that as an object starts to move, a Doppler mass center of gravity shift occurs which compresses the mass on the right side and has less effect on the far left. This counterbalances the effect of the acting force and the steady state solution becomes:

$$F_A > [\ldots\ldots\ldots\ldots\ldots\ast\ldots\ldots\ldots\ldots\ldots] < F_D \qquad \text{Figure 7-2}$$

In Figure 7-2, the applied force F_A is counteracted by the Doppler force F_D. Thus we get the inertial law of physics:

INERTIAL LAW 1: AN EQUAL AND OPPOSITE FORCE IS BUILT UP WHEN A FORCE IS APPLIED TO AN OBJECT.

We see that the applied force causes a physical compression to occur which reduces its atomic spacing and proton/neutron internal spacing as well in the object. The Doppler effect causes equal and opposite forces to occur. We have identified the reason for the inertial law from a space-time perspective. From the electrical perspective, the displacement of each electron, proton, neutron, and Bohr orbit from its image causes a counter-force due to Ampere's law of current loops.

We know that an object in free space continues on a straight-line course. Let us now remove the applied force. The center of gravity diagram now shifts to the right.

$$[\ldots\ldots\ldots\ldots\ldots\ast\ldots\ldots\ldots] \qquad \text{Figure 7-3}$$

In Figure 7-3, we no longer have any applied forces. There no longer is any counter force. However, we have a permanent shift in the center of gravity of the object to the right. Thus we have a permanent momentum and a permanent shrinkage of the object.

INERTIAL LAW 2: An object in motion will continue to move in a straight line because the center of gravity of the object is to the right of the actual physical center.

From the electrical perspective, we have each atom producing a momentum vector. This produces a spectrum of vectors all over the object. This maintains stability since the forces operate in a plane perpendicular to the motion. Every neutron and proton has their own little vector and the entire mass has billions of these vectors. The entire spectrum of vectors tends to prevent any turning motion. The center of gravity is the vector center but it is the individual vectors, which prevent the object from turning.

In the inertial law, we see that an object in motion will continue in motion and that it will be a straight line unless acted upon by a force.

Let us take our mass moving in the X direction with velocity V and approach a planet sitting in free space. The planet is exactly at rest with a complex spectrum gravitational field extending outward toward the radius of the universe. Our cubic mass has been shrunken in the X direction with a Doppler mass center of gravity to the right.

Let us put the mass above the planet as shown in figure 7-4 below:

$$M[\ldots\ldots\ldots*\ldots\ldots] \qquad \text{Figure 7-4}$$

P

The mass is moving to the right with velocity (V). It is above the planet in the Y direction. The gravitational field of the planet extends to the radius of the universe and exerts a downward pressure on the moving mass. The gravitational force does not pull on the mass. There is an outward pressure of the field from the radius of the universe backwards toward the planet. Thus gravity pushes objects together and does not pull objects together.

Notice that the center of gravity of the mass is to the right. The force of gravity is applied to the center of gravity and thus the object has an angular space time momentum applied to it. The object will turn slightly.

We see that if a moving object has a perpendicular force applied, it will turn because of the Doppler Mass shift. Our moon turns in perfect synchronization with us. It revolves around us and "locks in" due to the Doppler center of gravity shift. The Earth itself still has a lot of initial rotational energy and spins on its axis. Someday it will face the sun with only one face due to the Doppler Center of Gravity offset. Therefore in the far future, life on Earth will hardly be possible, as we know it.

We see that the circular or elliptical motion of the planets is due to the Doppler mass center of gravity shift, which causes the planet to follow the gravitational field of the sun.

INERTIAL LAW 3: An object moving in a straight line in the X direction and acted upon by a perpendicular force in the Y direction will move in a plane perpendicular to the Z axis. The velocity vector and the gravitational force vector form the plane that the object will move in.

In order for the object to reach stability, the following equation applies:

$$(MV^2)/R = GMM/R^2 \qquad (7\text{-}23)$$

This standard classical physics equation shows that the centrifugal force, (MV^2/R) due to velocity and curvature equals the centripetal force, (GMM/R^2) due to the gravitational field.

To see why the object curves, let us look at the center of gravity of the object in space-time.

```
          xxxxxxxxxxxxxx

          xxxxxxxxxxxxxx

          xxxxxxxx*xxxxx

          xxxxxxxxxxxxxx .........V        Figure 7-5

          xxxxxxxxxxxxxx

          xxxxxxxxxxxxxx

          xxxxxxxxxxxxxx
```

The X's represent the object which is moving to the right with the velocity V. The (*) represents the center of gravity of the object. Doppler velocity in the X direction causes the space time shift of the center of gravity to the right. This would cause straight-line motion. The force of gravity acting down on the mass causes a space-time shift of the center of gravity upward. The final center of gravity is shifted to the right and also upward in the Y direction. The velocity vector is shifted upward above the geometric center and the object moves in a circular path rather than a straight line. Thus the centrifugal force is due to the space time shift of the center of gravity of the mass.

It should be noted in the analysis that the individual Doppler shift for each proton is less than R_P. The majority of the measurable shift occurs in atomic spacing. Thus the actual shifts are quite small and complex from the simplification shown above.

SECTION 7-3: THE DOPPLER LENGTH

Now let us look at the details of the Doppler length that was presented in Chapter 6 section 5. This would be the length associated with the photon energy that is part of the inertial mass.

The inertial mass is the RMS Doppler mass and is centered forward of the moving object. The gravitational mass is contained within the object. The inertial mass tends to be outside of the object. This will be clear when we look at the Doppler length or inertial length. The gravitational length L_X is related to L_O by the equation:

$$L_{Xg} = L_O [1 - (V/C)^2]^{1/2} \tag{7-24}$$

In equation 7-24, the gravitational length of the X direction L_{Xg} as compared to the original length L_O shows shrinkage. The distance L_{Xg} would tend toward zero as the velocity increases. The distance L_Y tends to remain reasonably constant at the same time. We then have the following equations:

$$L_{Xi} = L_{Xg} / [1 - (V/C)^2]^{1/2} \tag{7-25}$$

$$\text{Doppler Length Front} = L_{Xg} [C / (C-V)] \tag{7-26}$$

$$\text{Doppler Length Rear} = L_{Xg} [C / (C+V)] \tag{7-27}$$

Let us now make a chart of this. It is important to see how far forward a mass actually is. This will aid in the understanding of the Double slit experiment, which will be explained in Chapter 9. It also helps us to understand the problems of high-speed space travel as explained in chapter 18.

V/C	Grav. L_{Xg}	Doppler L_X Rear	Doppler L_X Front	Inertial Length
0	1.000	1.000	1.000	1.000
0.1	0.994987	0.90434	1.10554	1.0000
0.2	0.979796	0.816497	1.22475	1.0000
0.5	0.866025	0.577350	1.73210	1.000
0.9	0.435890	0.229416	4.35890	1.000
0.99	0.141067	0.070890	14.1067	1.000
0.999	0.044710	0.022366	44.710	1.000
0.9999	0.014142	0.00707125	141.418	1.000
0.99999	0.004472	0.002236	447.20	1.000

In the chart we see that although an object shrinks to 4.5% of its original size when it achieves a speed of 0.999C, it has a forward inertial length of over 44.7 times its original size. As we head toward the speed of light, the object will approach infinite length. In the cyclotron, the forward Doppler length of the electron or the proton circles completely around the electron over and over again. The electron is turned into a massive circle of light and a strong Doppler image is produced.

The Doppler inertial length is the root mean square of the forward Doppler length and the rearward Doppler length. This equals the original length. However, the most important feature is the forward Doppler length. The same thing happens with Doppler masses. We get the forward mass and the rearward mass. With the forward length we see that an object's inertial image is ahead of its gravitational image. Thus an object arrives at a destination before it actually gets there. This will be a main factor in the double slit experiment explanation as well.

SECTION 7-4: CLASSICAL ORBITAL SPACE TIME MASS INCREASE

Let us understand the increase of orbital mass with respect to velocity from a standard classical physics perspective. Einstein's solution is a particular steady state electrical solution that includes cycle upon cycle of moving standing waves, which fill the entire orbit. The electrical equations are very simple and fit in perfectly with everything else. In integral calculus many times a solution is not readily available. People guess the solution and tables are made of solutions that work. The same is true for orbital space-time. It would be very difficult to calculate an orbital steady state solution. However we can see the ingredients of the solution.

Let us look at an object in orbital motion such as the electron or the Earth. The Doppler length to the right and left is:

$$L_{XR} = [L_O [1-(V/C)^2]^{1/2}] \cdot [C/(C-V)] \qquad (7\text{-}28)$$

$$L_{XL} = [L_O [1-(V/C)^2]^{1/2}] \cdot [C/(C+V)] \qquad (7\text{-}29)$$

Let us now look at the energy of the Earth moving in orbit.

$$E = \tfrac{1}{2} M_{OE} V^2 + \tfrac{1}{2} I \omega^2 \qquad (7\text{-}30)$$

In Equation 7-30 We see that the energy of the Earth in orbit is equal to its kinetic energy of $\tfrac{1}{2} MV^2$ plus the energy of rotation $\tfrac{1}{2} I \omega^2$. Normally the rotational energy is not considered. However as we speed the Earth up toward light speed, the kinetic energy will not be much. The best we can get is:

$$E = \tfrac{1}{2} M_{OE} C^2 \qquad (7\text{-}31)$$

As we achieve light speed we can get an equivalent amount of energy due to distortions. However Equation 7-31 will never get us to the huge amount of orbital energy within the cyclotron. It will not approximate Einstein's solution except at very low speeds.

Let us look at the second half of the equation. At first glance the radian frequency ω will not do better. We will never achieve Einstein's orbital equation from that. Let us now look at the equation for inertia. In standard classical physics, the moment of inertia of a hollow cylinder about its axis is:

$$I = MR^2 \qquad (7\text{-}32)$$

In equation 7-32 we see that the moment of inertia increases as the square of the radius. The moment of inertia of the electron increases as the square of the radius of the Bohr orbit and the moment of inertia of the Earth increases as the square of the distance to the sun.

As we look at equation 7-32 we see that the Doppler distance in the direction of motion rises toward infinity as we approach the speed of light. Thus the inertial energy rises toward infinity as we head toward light speed.

We now see the reason why Einstein is basically correct in his orbital energy equations. We are dealing with tremendous increases in the moment of inertia of the electron in the cyclotron. We are also dealing with more modest Einsteinian corrections in planetary motion, the Bohr orbit, and neutron calculations.

In conclusion, classical physics with Doppler space-time is the correct solution in all cases. Einstein's orbital solution is quite correct but it is a specific solution and not the general properties of space and time. The true equations of space and time are the GG/MM/Doppler equations with ordinary classical physics and corrections for increases of inertial energy with respect to velocity.

CHAPTER 8 ROTATIONAL SPACE TIME

SECTION 8-0: INTRODUCTION

In this chapter, rotational motion will be investigated for important concepts. The toy gyroscope will be explained. Finally a very constrained orbital clock in an airplane will be looked at. This will enable the reader to see the application of Doppler space-time and M/M space-time concepts applied to rotational motion. Only a brief treatment of rotational motion will be undertaken in this book.

SECTION 8-1: ROTATION IN M/M SPACE TIME

Let us now look at rotation in GG/MM/Doppler space-time. In Chapter 7 we saw that in the understanding of GG/M/M space-time, the dimensions of an object changed as the object was moved. Doppler space-time defines the change in the center of mass of the object as if it moves under linear forces and also under linear and perpendicular gravitational forces. It also shows the differences between the gravitational mass and the inertial mass. In addition, the differences between the gravitational length and the inertial length were shown.

When we look at an object in motion, the center of gravity is shifted in the direction of motion. We can look at this as if another object exists in front of the motion. The object is fast-forwarded in the direction of motion. In effect, an object precedes its arrival. This is similar to an Einsteinian concept in which space appears shortened to something moving at the speed of light.

The Doppler effect causes this phenomenon. The wave front in front of the object travels at the speed of light C. and the object travels at V. The wave front or Doppler mass arrives before the object. If we look at a train with a searchlight traveling upon a circular track, you will see the light first at night, and then you will feel the train. The train actually arrived prior to it's own arrival. The Doppler mass wave front reached you before the train. This is one explanation for the double slit experimental effects that will be explained in Chapter 9.

If you took a huge star and brought it up to the speed of light, its effects would be quite tremendous far from its actual location. You could be smashed by the forward image of a star millions of miles before it actually hits you. If you look at the electron in the cyclotron, the Doppler wave front is initially ahead of the electron as it starts to move. In a short time, the wave front will circle the cyclotron and it will catch up to the electron. Then it will pass it. This will continue over and over until the electron is moving at the speed of light. You end up with a combination photon/electron, which is a line of charge. An image is also produced of the electron, and the electron is everywhere.

Let us now look at a simple shaft that is rotating clockwise. Let us look at one atom within the shaft. We can place one radioactive atom within the shaft as a marker. Let us call this zero degrees. As we rotate the shaft, the space-time projection of the shaft moves clockwise. The Doppler shift in the center of mass for linear motion becomes a Doppler shift in the phase angle or torque angle for rotational motion. Changes will occur in the shaft radius, which will be discussed later. For the moment we have two objects. We have the original shaft and a ghost shaft. The ghost shaft moves ahead of the original shaft. It can move 360 degrees. This will be a phase locked condition.

We see that the energy of rotation is stored in the ghost inertial mass. The more we rotate the object, the greater will be the phase shift. Under normal circumstances, if we rotate close to near light-speed its gravitational energy will double. Rotational energy is very similar to Einsteinian orbital energy. You can build up quite a lot of energy. However physical limits tend to be reached.

In the rotational system, things are quite confined. If we double the mass/energy, we can cause the shaft to fill in the image of itself in space-time. This tends to be a natural limit and it tends to be only a gravitational self-image because beyond this point it may explode.

In rotational systems the original object will tend to shrink in the axial direction. This is all gravitational energy. Simultaneously inertial energy will extend out from the object. This heads toward infinity. This is very similar to a magnetic field. As each atom tends toward light speed, its linear projection arm heads toward infinity. The inertial distance of each atom enlarges in a similar way as with linear motion.

We then get an inertial gravitational field as part of the object. The energy of rotation is kept both in the Doppler inertial mass phase shift within the object and within the inertial gravitational field of the object. These two ingredients turn rotational motion into Einsteinian type motion. An object moving in free space does not have an inertial gravitational repetitive image. It can achieve light speed whereas the rotating device will start to get repetitive standing waves of rotational energy, which makes it quite massive. We can rotate it at near light speed with tremendous mass/energy.

It is the rotational Doppler effects, which produce the galaxies. We have an inner gravitational field of the galaxies and an outer inertial (magnetic type) rotating gravitational field. We can have galaxies that turn from a perfect sphere into a flat disk. There are always acceleration terms so the Z-axis or rotational axis will vary. In linear space-time, we looked at invariant Y and Z. There will be some variations in Y and Z especially at the light speed barrier. However linear space-time does not have the constant acceleration forces upon it. You can shrink the X direction pretty far before you hit non-linear effects. This is not easy to do with rotational motion since the centrifugal forces are always putting the object under great stress. Pure dots can do much more than ordinary atomic structures. You can get a near light speed disk of pure dot energy, which contracts to form a black hole, and then the black hole later explodes to form a galaxy.

Let us now write some rules for rotational Doppler effects:

ROTATIONAL DOPPLER RULES:

1. An object rotating in the clockwise direction will have a Doppler shift in the clockwise direction. This can be defined as a phase angle or torque angle of the mass. It will be a ghost mass.

2. An object rotating in the counterclockwise direction will have a Doppler phase angle shift in the counterclockwise direction.

3. If you bring an object up to approximately 86.6 percent of the speed of light, it will achieve a 360-degree rotation and will be phase locked. Its mass/energy will be twice that of the original rest mass of the object. Its inertia will be twice as much also.

The rules of rotational Doppler are simple enough. However, once we start to rotate an object we can get various changes in the radius and the shaft length. This will now be discussed.

Let us look at a cylindrical object in rotation of 1-meter radius (R_X) and 1 meter length (L_Z). We find that we no longer have the same amount of choices as with a cubic block of mass with three separate dimensions. If we rotate the object around the Z-axis, we can no longer have any differences in X and Y. Of course we could always move the entire structure in the X or Y direction or at an angle to both which would produce complex variations. However, let us only look at simple axial rotation in this section.

For simple axial rotation, we only have the radius dimension (R) and the Z direction. The circumference is looked at as the ($2\pi R$) dimension that tracks the radius dimension.

In our ordinary experiences, if we take an object such as a grinding wheel we see that is contains a maximum rotational speed rating. The reason for this is that there will be a physical tendency for it to fly apart. The centrifugal forces within an object tend to enlarge it as it spins rapidly and the centripetal forces within the object tend to shrink it or hold it together.

If we consider one tiny cube of material within the spinning object, it will shrink in the ($2\pi R$) direction caused by the rotational motion perpendicular to the Z-axis of rotation. It will enlarge in the R direction due to the centrifugal force, which also causes the ($2\pi R$) direction to enlarge, and the Z direction to shrink.

Of course it is not possible for an object to shrink in the circumference dimension without actual shrinkage in the radius dimension. A rotating object shrinks due to Doppler space-time centripetal forces and expands due to ordinary centrifugal forces. We then get two solutions. The first is when the centrifugal forces are the leading forces and the second is when the centripetal forces are the leading forces. Let us look at a situation where the centrifugal forces are the leading forces such as in an expanding galaxy. In this case R increases and Z decreases. Thus:

$$R_X > 1 \tag{8-1}$$

$$L_Z < 1 \tag{8-2}$$

If we bring the galaxy up toward the speed of light, it will tend to reach toward infinity in radius and reach zero thickness. In this case the centrifugal forces are driving the expansion. We see that a small round ball galaxy can become a large flat disk.

Let us now take a galaxy in which the forces are such that the Doppler centripetal forces are stronger than the centrifugal forces. The galaxy shrinks in diameter and expands in length or thickness as it slows.

Galaxies have only gravitational forces acting upon them. They will behave somewhat different from that of a rotating steel rod. The steel rod has strong binding energy forces acting upon it.

The solution to the rotational problem is to produce a cylinder or sphere out of many tiny cubic boxes of matter. The only difference is that all the boxes are tied together by the binding energy of the material. You can take an iron sphere and bring it up in linear motion toward the speed of light. Yet, if you took the same iron sphere and rotated it pretty fast, it would fly apart rapidly. The centrifugal forces tend to build up very strongly and rip the object apart. Of course you would do better with an electron or proton. You should be able to spin them pretty fast because they have little mass and very strong electrical binding forces. They will tend to turn into a flat disk but depending upon the experiment they could turn into an inner line and an outer inertial field.

If we try to spin a galaxy too fast, it will just reach a high speed and then start to fly apart because the gravitational forces are just not that strong to hold it together. The only thing you could spin at the speed of light would be the primary dots themselves. A large flat plane of dots could rotate at the speed of light and when it slows could form a super neutron the size of this Earth. In this case we are not dealing with the weak gravitational forces of attraction but the extremely strong primary coulomb forces of the dots themselves.

A galaxy of primary free space dots could merge together and form a high-speed rotating disk that eventually would slow and merge into stars, solar systems, and planets.

Once the object no longer is a singular electrical entity such as a super neutron, the centrifugal forces could take over and drive the galaxy apart. Yet, each problem can be different as we are dealing with a delicate balance of forces that could go either way.

Let us now look at the Acceleration Doppler shift for an inertial object. In basic physics, the moment of inertia of a solid circular rod around the axis is:

$$\text{M.I. Rod} = \tfrac{1}{2} MR^2 \qquad (8\text{-}3)$$

We could replace this rod by a mass at $0.707R$, which would be the inertial center of gravity (ICG). If we rotate the rod, there will be a physical force expanding the rod as it rotates faster and faster. There will be a Doppler counter-force tending to constrain the motion as the equal and opposite force. Finally when the torque is removed, we are left with an Acceleration Doppler ICG shift. This is in addition to the normal Doppler velocity shift in phase angle around the direction of rotation. The acceleration shift is show in Figures 8-1 and 8-2.

    o[xxxxxxxxxxxxxxxxxxxxxxxxIxxx*xxx]   Fig. 8-1

In Figure 8-1, the right cross section of the rod is rotating about the center of the rod shown as (o). The ICG is shown as (I) and the shift in the ICG due to centrifugal force is shown by the star (*) to the right of the (I).

A Doppler counter force is built up which tends to shrink the ICG of the rotating object to the left of the ICG. This shift in ICG due to the Doppler counter force alone is shown in Figure 8-2.

    o[xxxxxxxxxxxxxxxxxxx*xxxIxxxxxx]   Fig. 8-2

In figure 8-2 we see the Doppler counter force for an inertial motion. There is a constant acceleration of rotational motion that makes the problem different from simple linear motion. A centrifugal force will always exist within the rotating object. The Doppler rotational centripetal force will always act to counter this. However, for most objects there will be a permanent expansion of the radius. The ICG is permanently shifted to the right for masses in pure free space. The kinetic energy of rotation is maintained by the phase angle of the rotational Doppler shift.

We see in rotational motion, the combination of rotational Doppler, centripetal forces, centrifugal forces, binding forces and internal heat forces which tend to change the balance of the two accelerating forces. If we add additional forces such as magnetic and electric fields, then we can change the shape of the rotating disk.

For most cases as we rotate the mechanical objects faster and faster, the thickness or Z dimension of the rod shrinks toward zero. The same would be true if we started with a sphere. As we move faster, we get a flat ring surface. As we move even faster toward the speed of light for electrostatic bonding rather than gravitational bonding, we get a loss of the center of the ring. Thus we get a washer effect.

What we have produced by the shift in the ICG and the flattening in the Z dimension are the Rings of Saturn. The rings must be charged particles so that they have the strong coulomb attractive force rather than the weak gravitational force, which would not hold them together. Primary dots would work as well but the rings of Saturn appear to be charged particles such as ions. If we move toward the speed of light, the disk will change into a large circle in a flat plane.

In rotational motion we have conflicts between the Doppler velocity vector which tends to keep the object flowing in a straight line and the Acceleration Doppler vector which tends to shrink the object. This is counter balanced by the centrifugal forces. When the velocity is slow compared to the speed of light the object can maintain its radius since the ghost-rotating shaft fits within the object. The linear shrinkage of the object tends to exert the axial inward centripetal force. However, the ghost can be considered smaller than the original shaft. This causes an outward pressure or centrifugal force.

We then have two conflicting space-time acceleration problems. In the rotating shaft there is little place to go. As we speed the object up faster and faster we have a nonlinear problem rather than a simple linear problem. The energy coming into the object can reach twice the original mass/energy. After that point the inner ghost can be considered a series of inner ghosts which are smaller and smaller. This causes tremendous pressure. Thus the ability of the object to shrink slightly at low rotation is negated once all that energy starts to flow into it.

The net result is that the object starts to expand rapidly as we approach the speed of light. Of course most objects fly apart long before that. However, this is applicable to a neutron star, which already has shrunk down to a tiny size. As more and more photon energy comes into it and we reach twice the mass/energy of the original neutron star, then there is no ability to contain it anymore.

If you look inside the neutron in that star, the energy is compressed spherically into a little ghost of the neutron. There is no place for this energy to go. You do not have any ability for it to move or elongate in any direction. It will initially shrink as energy is poured in. The ghost neutron within the neutron puts tremendous pressure on the neutron.

The neutron star will explode. Black holes will also explode. And the entire universe when compressed far enough will produce a space-time ghost universe. This universe would be approximately half the volume of a pure neutron universe. This dimension can be calculated from the number of neutrons in the universe. Of course everything shrunk and the ruler is smaller, but relatively we can calculate readily the exact size of the universe at big bang. This produces a slightly different answer than the calculation shown in Chapter 3.

It is interesting to note that since we must compress the universe to half the volume in order to make it explode, there must be twice the amount of energy in the universe. This will be found in free dots and photon energy to account for the loss of an equal amount of energy at big bang. This does not change the oscillating cycle time of the universe, which was derived from electrical characteristics. It will enable astronomers to understand that in order for the big bang to have occurred, twice the energy was necessary at the time.

Just as a moving object has twice the gravitational mass/energy at light speed, the universe needs twice the mass/energy in order to explode. This is an important conclusion.

SECTION 8-2: THE GYROSCOPE IN SPACE TIME

Let us look at a simple gyroscope and see how it works. The Author experimented with a simple toy gyroscope. Nowadays they come in half metal/half plastic, which makes them much more pleasant than the old all metal types. The purpose of the gyroscope experiment is to learn about rotational motion. Anyone can do the simple experiments themselves.

The gyroscope is handy because a spinning gyroscope that is spinning in a vertical plane and pivoted at one support end will rotate clockwise or counterclockwise when looking downward on the table. This will help assist the Author and reader in understanding the Doppler center of gravity shift.

It appears that when we push a mass in free space, it will compress where we push it. This will produce a higher density of mass on the side where the force is applied. Let us push the object on the left. It will compress on the left side. The center of mass will shift to the left. The mass will move and there will be a Doppler mass shift on the right that is almost equal to that on the left. When the force is removed, the compression on the left and the right reduce but the Doppler shift remains on the right side.

Now if we take an object in rotational motion, and look at a tiny piece of it, it will tend to shrink in the direction of motion. If we take a gyroscope wheel and rotate it, the circumference at every radius from the center will tend to shrink. There will be force acting toward the center of the gyroscope due to space-time shrinkage. This is counterbalanced by the centrifugal force that tends to send the parts of the gyroscope wheel flying outward.

In the case of circular motion, the space-time effect tends to act like a gravitational field producing a force directed from the Earth toward the sun. If we use a string on a weight and swing it, the string will act the same as the gravitational force does. Yet, as we look at the weight as we spin it very rapidly, there will be shrinkage of the circle closest to the axis of the string. Thus the space-time forces will still act toward the center of the rotation.

Returning to the gyroscope. Let us have a gyroscope in the vertical X/Y plane spinning clockwise. We will have a Doppler mass shift to the right at the top of the wheel. We will have a Doppler mass shift to the left at the bottom of the wheel. We will also have a Doppler mass shift to the top at the left of the wheel. Finally we will have a Doppler mass shift to the bottom to the right of the wheel.

This is shown in Figure 8-3:

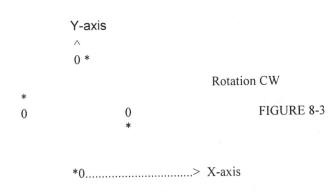

FIGURE 8-3

In Figure 8-3, the (0) represents the top, bottom, left, and right of the gyroscope wheel. The wheel is moving clockwise. The (*) represents the Doppler shift at the quadrants. The momentum of the wheel is maintained by the Doppler shift of the masses. There will be a force of acceleration from every quadrant to the center. At the top, there will be both a force of acceleration and a force of gravity. At the bottom, the force of gravity will be slightly stronger than at the top.

We see that the acceleration of the atoms of the gyroscope on the left side and the right side will produce equal forces in the downward direction. There will be no net torque for atoms on the left side or the right side since the Doppler shift in the center of gravity (C.G.) is balanced as per the center of the gyroscope. In addition, the forces of gravity are equal in the horizontal plane.

When we look at atoms at the top and bottom of the gyroscope we see that the Doppler shift in the center of gravity are equal and opposite. However, since the acceleration due to gravity is stronger on the bottom than the top, a net torque occurs. When the gyroscope moves off a perfect vertical plane, the net torque forces the gyroscope to precess counterclockwise. If we look downward upon the gyroscope at the X/Z plane, we see that it rotates counterclockwise.

If we change the rotation of the gyroscope to the counterclockwise direction, the unbalanced torque will be in the opposite direction and the gyroscope will precess clockwise. The study of the gyroscope produces several space-time conclusions for rotational motion:

Law #1: An object in rotational motion moving clockwise has a Doppler mass shift at an angle which varies with velocity and which is shifted to the clockwise direction. The Doppler shift is always in the direction of rotation ahead in time of the object itself.

Law #2: An object in rotational motion tends to shrink in circumference at every point and has force acting toward the center of rotation. This is a Doppler shift centripetal force that is similar to a gravitational force for planetary motion. This force is balanced by the centrifugal force acting outward from the direction of the center of rotation. In addition, there are internal binding forces acting and internal photon expansion forces acting. In general mechanical structures will expand in the plane of rotation during the spin, while electrical structures can be made to shrink. Each problem must be analyzed for all the forces involved.

Law#3: An object in rotational motion and with a horizontal axis, and spinning clockwise from the pivot point will move counterclockwise when looking downward. This is due to stronger gravitational forces of attraction at the lower side than at the upper side. In normal physics both the centrifugal and centripetal forces would be balanced out and it would be hard to see an unbalance. However, in space-time physics, the acceleration pulling the gyroscope material to the axis, shifts the center of gravity toward the axis and this does not balance out with the centrifugal force. This results in a rotational space-time vector, which is part of the Doppler shift.

If we look at each piece of the gyroscope, we see that rotational motion is really little pieces of linear motion added together. We get a Doppler shift which can be considered the same as a phase angle. However this is basically the same as the linear Doppler shift except that the acceleration changes the momentum vectors. When we pull down on a Doppler shifted C.G. we change the horizontal value of the momentum vector. The vectors do not add linearly since we are working at the speed of light. Thus by increasing the downward force and vector, we reduce the horizontal vector. This is another way at looking at the unbalance within the gyroscope.

SECTION 8-3: ORBITAL CLOCKS

Now let us look at a clock orbiting the Earth. This tends to be the Einsteinian orbital clock solution. This clock is constrained by the Earth's gravitational field. It is not a free clock. All the clock solutions tend to be somewhat different. We have to find the boundary conditions that constrain the clock. In the ideal free clocks for linear space-time we set physical boundary conditions such that the Michelson/Morley apparatus nulled at all times.

There are many solutions possible in which the apparatus might null in two axes and not null when the third axis is included. We could also look at solutions where you can null it at zero absolute velocity and it will never null again. Then it becomes an absolute velocity detector. All these different solutions require experimental data. In this book only the GG/MM/Doppler solution is discussed. This is one particular space time solution which brings agreement between classical physics and Einstein's orbital physics.

We live on planet Earth. Our measurements and physics are based upon planet Earth. If we lived on a high velocity planet elsewhere, we would have to use different describing functions to explain the universe. If we maintained volumetric integrity, we get different space-time solutions. We can have a full set of space-time equations such as:

$$L_X L_Y L_Z = L_O^3 \quad (8\text{-}4)$$

$$T_X T_Y T_Z = T_O^3 \quad (8\text{-}5)$$

$$C_X C_Y C_Z = C_O^3 \quad (8\text{-}6)$$

Equations 8-4 through 8-6 describe a universe that maintains volumetric integrity. Distances and times and light speeds can vary. We get a universe that can be a straight line with infinite light speed in one direction and zero light speed in the other directions. This can become an elliptical plane, and later it becomes a circular plane. Then we have a two dimensional universe. Later it could start to become an ellipsoid and finally we end up with a perfect spherical shape.

We can produce space-time equations and space time clocks for a whole variety of universes. Even within our universe, there are many non-linear areas, which will never null the Michelson/Morley apparatus. All we can do is to note that beyond our Earth lie an amazing variety of things.

When you are in free space, there is no relationship between the Doppler frequency to the front and the Doppler to the rear that constricts you. When you circle the Earth or any other orbit, you are limited by the rules of electrical theory. In orbit for steady state motion, you are bound to the rules for phase locked loops.

A phase locked loop will lock on a particular frequency. It will equalize the frequencies. If we take a clock of ground frequency f_O and place it in an airplane circling the Earth, the Doppler frequencies are:

$$f_{O\ Left}' = f_O\ [C/(C+V)] \qquad (8\text{-}7)$$

$$f_{O\ Right}' = f_O\ [C/(C-V)] \qquad (8\text{-}8)$$

$$f_{O\ RMS}' = f_O\ /\ [1-(V/C)^2]^{1/2} \qquad (8\text{-}9)$$

The geometric mean or root mean square Doppler is equal to the inverse of Einstein's formula. If f_O remained constant at the ground frequency, it would be operating at a higher energy level in the Airplane than on the ground. The airplane itself is certainly operating at a higher energy level by the increase in kinetic energy. However, the clock is being dragged large distances which means its time of oscillations are related to the aircraft velocity and the speed of light. The solution is that the clock drops frequency so that for a constant Doppler time solution we get:

$$f_O'' = f_O\ [1-(V/C)^2]^{1/2} \qquad (8\text{-}10)$$

In equation 8-10 we see that the clock was forced to drop frequency to match the Doppler RMS value. If the clock maintained its original frequency, it would not match a phase lock loop steady state electrical solution. The true clock speed for the Doppler universe is:

$$f_T = f_O \qquad (8\text{-}11)$$

In equation 8-11 we see that the true clock is identical with the clock when sitting on the ground. This clock is a RMS Doppler clock. It takes into account the RMS Doppler time changes. It is a more sophisticated clock than a normal clock. For the universe:

$$\text{RMS Doppler time} = \text{Invariant} \qquad (8\text{-}12)$$

When you travel space and time you need a true invariant clock. The true time clock needs a computer as part of it. It needs an absolute velocity sensor as well. This is a fancy clock but this clock will read identical time everywhere in the linear universe. Once you get into a black hole or ultra-high speed planet in a very non-linear area of space-time, the clock will not give good results.

Let us have an airplane flying with velocity V in a straight line. Let us compare the clock rate of a standard typical clock with a point on the ground. Since the space-time distortions of the Earth as it moves in the Universe tend to be common mode to the space-time distortions within the Airplane, they cancel out. The only thing important is the airplane itself. It really does not matter whether the Earth is moving or not when you compare a point on the ground with the airplane. There may be some differences between the equator and the North Pole since the ground clocks will be different. However, once we choose a reference point, all the changes with respect to frequency become common to the reference point and the airplane flying overhead.

The airplane clock is oscillating at the frequency fo when stationary on the ground. The airplane will take off and circle the Earth. The movement of the clock with velocity V relative to the sphere of the Earth will produce both a Doppler to the right and a Doppler to the left. The Doppler fields travel around the world at the speed of light C. When the plane moves they will circle the Earth over and over again. In a short time a steady state field exists. This will tend to reduce the clock speed in the direction of motion. The frequency must drop for a phase locked condition.

Let us assume that (S) is the circumference of the Earth and that (V) is the velocity of the Airplane with respect to the surface of the Earth. The clock will put out a burst pulse every time it crosses the zero volt electrical point. We have a particular frequency with a tiny burst pulse readily available for time measurements.

The pulse will travel the circumference of the Earth and hit the back of the Airplane's clock. It will rebound for the steady state solution and return to the front of the clock after traveling the Earth. The time of flight is:

$$T_X = S/(C+V) + S/(C-V) = (2SC)/(C^2 - V^2) \quad (8\text{-}13)$$

$$T_X = (2S/C) / [\,[1-(V/C)^2]^{1/2} \cdot [1-(V/C)^2]^{1/2}\,] \quad (8\text{-}14)$$

We see in equation 8-14 that the time in the X direction is reduced by the square of the Einsteinian correction factor.

Now let us look at the perpendicular direction. The pulses from the clock move perpendicular across the surface of the Earth. They travel at light speed all over the Earth and meet up again with the clock. The time for the round trip is:

$$T_Y = (2S/C) / [1 - (V/C)^2]^{1/2} \quad (8\text{-}15)$$

We now have two times which are unequal. The frequency of the clock is determined by the Y direction measurement, which is easier to obtain since it passes an instrument on the ground and is free of the Doppler effect. The clock reads:

$$f_O' = f_O\,[1-(V/C)^2]^{1/2} \quad (8\text{-}16)$$

We see that the clock frequency has been reduced by the Einsteinian factor. If the speed of the airplane reaches light-speed, the clock will stop.

In the X direction the clock frequency must be identical as in the Y direction. The Y direction slows the clock because the clock keeps moving and the distance traveled within the clock keeps getting larger due to the velocity of the airplane verses the speed of light. This slows the clock. This is merely classical physics electrical theory.

In the X direction, the time is more than the Y direction by the Einsteinian factor again. For stability the pulse on the wave at the zero crossing must exactly match the pulse on the wave in the Y direction. The Michelson/Morley experiment brings these two pulses together simultaneously. That was before the days of Doppler radar and phase locked loop theory.

When we look at the aircraft clock we see a higher frequency coming out of the front of the plane. This circles the Earth. It meets up with a lower frequency coming from the rear of the plane. Halfway around the world you end up with a junction of this frequency modulated wave that is exactly the same as the frequency of the airplane's clock. This is the Doppler image of the clock.

Looking at the rear of the plane, you see lower frequency waves coming out and moving around the Earth and meeting up with the airplane's frontal waves. In order for a steady state electrical solution, the FM modulated waves from the back must match the FM modulated waves from the front. We get a very fancy standing wave pattern that moves. The airplane carries its own image around the Earth and it also carries the FM modulated clock.

There is no physical shrinkage of the clock. The frequency must drop so that the number of cycles of the clock in the X direction exactly equals the number of cycles in the Y direction. The little cycle times are all different. In addition the number of cycles in the steady state solution is different in the X and Y-axis. Initially for the transient solution, things are different. It is more like linear space-time.

However, as the waves travel the Earth approximately four times, we start to get phase locked FM waves. The alternative to both Einsteinian space-time and GG/MM/Doppler space-time is FM Doppler space-time, which will be discussed in Chapter 11. In FM Doppler space-time we live in a frequency modulated universe. No variation of X and Y occur. Everything is based upon phase time. Sine waves get larger and smaller and adjust.

We see that even with the clock problem, standard classical physics using variable FM modulated waves eliminate the need for huge amounts of energy to travel the stars. The GG/MM/Doppler space-time does not require this, nor does the FM Doppler universe.

Let us now return to the clock problem using the approximations for low velocities as compared to the speed of light. The frequency change of the Airplanes clock for the Einsteinian/Doppler solution at low speeds compared to the speed of light is:

$$f / f_o = 0.5(V/C)^2 \qquad (8\text{-}17)$$

The time loss for a plane orbiting the Earth for the Einsteinian solution is:

$$\text{Time} = 0.5[(V/C)^2](S/V) = (1/2)(V/C)(S/C) \qquad (8\text{-}18)$$

The time solution depends upon the Doppler ratio V/C and the time it takes to orbit the circumference of the Earth S/C where $S = 2\pi R_E$.

Equation 8-18 [Time = (1/2)(V/C)(S/C)], where V is the velocity of a plane orbiting the Earth, S is the circumference of the Earth, and C is the velocity of light gives the time clock error as per Einstein's derivation.

Let us examine the time lost in an airplane flight or satellite flight around the Earth. The change in clock frequency is:

$$f / f_o = (1/2)(V/C)^2 \qquad (8\text{-}19)$$

The time to orbit the Earth is:

$$T = S/V \qquad (8\text{-}20)$$

Where S is the circumference of the Earth, thus:

$$S = 2\pi R_E \qquad (8\text{-}21)$$

The radius of the Earth is the radius at satellite orbit or airplane height plus earth's radius. The loss of time is:

$$\text{delta Time} = (1/2)(V/C)^2 (2\pi R_E)/V \qquad (8\text{-}22)$$

Thus:

$$\text{delta Time} = (V/C)(\pi R_E/C) \tag{8-23}$$

Equation 8-23 is a very important equation. It shows that the loss of clock time is equal to the ratio of the satellite speed to that of the speed of light (V/C) and the time to circle half the globe by a beam of light ($\pi R_E/C$). Since V/C is the Doppler ratio, it shows that the space time problem for orbital motion is the Doppler and that Einstein's formula and the Doppler RMS are identical.

Let us return to the time formula. We know that the Doppler to the right and left are:

$$f_{Right} = f_O"(1+V/C) \tag{8-24}$$

$$f_{Left} = f_O"(1-V/C) \tag{8-25}$$

The average Doppler is:

$$f_{Average} = f_O" \tag{8-26}$$

If we look at the Doppler patterns in the direction of motion all around the Earth, we will find that they produce standing waves within the gravitational field. The oscillating crystal produces waves to the left and waves to the right, which meet halfway around the Earth.

This means that halfway around the Earth a virtual clock exists which operates at the frequency $f_O"$.

The same thing happens in the Bohr orbit, where there is a dual electron halfway around the orbit. This causes the repulsive electrical forces when looked at as an electrostatic problem. The answer is the same when we view the problem by the current formula for the repulsive electron current loop. The dual electron is a ghost or shadow electron. It is a low energy image.

The interesting thing is that the steady state solution produces a dual Earth as well. In the case of the airplane's clock, the problem is simple. The gravitational waves circle the Earth and form standing waves from both directions and which meet on the opposite side of the Earth. This is the null point, which looks like an image of the original clock. The dual Earth has limited energy since the speed of the Earth is low compared to the speed of light. However on the other side of the sun, the dual Doppler Earth does exist.

The virtual Earth is produced on the opposite side of the orbit. Since the velocity of the Earth is small compared to the cyclotron or the Bohr orbit, the virtual Earth is more of a low energy collective intelligence field rather than something solid that we can see and feel. However, a spaceship at the location of the dual Earth will find strong gravitational forces acting within the field.

The meaning of the existence of the dual Earth on the other side of the sun is for philosophers to ponder. We exist here but we also exist everywhere in the steady state Earth's orbit of over hundreds of millions of miles. In particular, there is a dual or ghost image of us with focal point upon the dual Earth. This is basically pure photon energy.

The mathematical analysis of the Doppler earth and our Doppler Image necessitates a philosophical understanding of the phenomenon. This is presented in Chapters 15-17 of this book from a scientific philosophical viewpoint.

CHAPTER 9: SPACE & SPACE-TIME

SECTION 9-0: INTRODUCTION

In this chapter we will investigate some of the properties of space and space-time. We will look at the question of whether there is an ether or not. We will look at the dot again. We will also look at the double slit experiment from two perspectives.

SECTION 9-1: SPACE ETHER

If we were born in the body of a Jellyfish and never ventured far from the center, we might conclude that the Jellyfish was made up of particles, which reacted in some strange way. As we studied our existence we could formulate rules and regulations concerning what was happening in our Jellyfish universe. We would never see the entire Jellyfish and scientists could argue back and forth over whether only particles existed or whether some ether existed.

In the same light, we live within our universe and do not see the entire thing. We can only speculate whether it is the dots themselves, which make up the entire universe or whether the dots are the property of the space ether. The space ether gives us a space time universe. It is a universe with a body.

The particles of our existence are hard. They are the centers of a huge number of dots arranged in a spherical pattern. The waves of our existence are soft. They are the mass-less inertial energy of photons and the magnetic fields. They appear to be perturbations in the ether.

The dot itself contains an inner part and a field, which extends all over the universe. We can then build the entire universe from the dots themselves. No ether is necessary because the dots are the ether.

If we go back in time long before the present universe, we can find a state of pure nothingness. Pure nothingness is something. Pure nothingness is a pure homogeneous electro-magnetic field. The state of the universe of pure nothingness is pure homogeneity. The homogeneous electro-magnetic field has no distinguishing features. A spaceship within this field can find nothing at all.

The origin of the universe can be traced back to an original pure state. We can look at the electro-magnetic field and see an infinity of configurations from the present universe, to pure nothingness, to a multi-light-speed set of universes, and to a universe of pure standing waves of energy.

We cannot go beyond the original pure state of nothingness. The substance of our bodies and minds originates within the electromagnetic field. All we can see and understand is the infinite ability of the field to produce an amazing assortment of universes.

In Chapter 14 we will look at a philosophical solution to the universe. We must always start with the physical fact that the electromagnetic field in one form or another always exists. Philosophers and theologians can ponder how the field originally came into being. Yet it is not really possible to understand such things since we only have the ability to change one form of energy into another form of energy. We can make a universe from pure nothingness because pure nothingness is merely a purely homogeneous state of something.

Let us now look at the primordial universe of pure nothingness. We can look at this state as the summation of two dots, a singular plus dot and a singular minus dot which are superimposed upon each other. The net result is that the electromagnetic field completely cancels out everywhere. We obtain an infinite universe where the electrical currents everywhere are zero. Thus:

$$I_U = 0 \tag{9-1}$$

One property of the primordial universe is that the current flow everywhere is zero. Mass requires current flow therefore:

$$M_U = 0 \tag{9-2}$$

In equation 9-2 we have a universe of absolutely zero mass. Without current flow, mass cannot be produced. Therefore the primordial universe is mass-less. The size of the universe is infinite. Thus:

$$R_U = \infty \tag{9-3}$$

In equation 9-3 we have a primordial universe which is infinite in extent. There is no ruler anywhere but space is infinite. The time clock of the universe is not moving. Therefore we have infinite time as well.

$$T_U = \infty \tag{9-4}$$

Both time and distance of the primordial universe are infinite. The light speed could be finite but it is infinite as well. Thus:

$$C_U = \infty \tag{9-5}$$

We have zero mass and infinite light speed. We have zero energy in the form of mass. We also have zero kinetic energy since nothing in the universe is moving. All that we have is potential energy. Therefore:

$$E_U = \infty \tag{9-6}$$

The energy of the primordial universe is infinite. However there is no current flow and therefore no mass. The primordial universe is a mass-less universe. The energy is all potential energy. The energy is in the form of charge depleted to zero at infinity but having the capability of returning to strength when the universe compresses. The same is true of the potential capacitance of the universe.

$$\underline{C}_U = \infty \tag{9-7}$$

In equation 9-7, the potential capacitance of the universe is infinity. In the present universe as we stretch out toward infinity, the capacitance drops toward zero. The same would be true for the primordial universe. However in equation 9-7 we look at the potential capacitance. This is similar to potential energy. The compression of the universe produces infinite capacitance. The expansion of the universe produces infinite potential capacitance.

The same is true of the charge of the universe. The charge at infinity is zero. We have plus charge and minus charge of zero. However the potential charge is infinite. Therefore:

$$+Q_U = +\infty \tag{9-8}$$

$$-Q_U = -\infty \tag{9-9}$$

$$\text{Net } Q_U = 0 \tag{9-10}$$

In equations 9-8 & 9-9, we have potential charge of infinity for the primordial universe. The net charge is zero as the plus charge counterbalances the minus charge.

We also have potential voltages. The voltages are zero everywhere in the primordial universe. Therefore:

$$+V_U = +\infty \tag{9-11}$$

$$-V_U = -\infty \tag{9-12}$$

$$\text{Net } V_U = 0 \tag{9-13}$$

The primordial universe of pure nothingness has infinite positive and negative potential charges, which are evenly, distributed everywhere. However, a test spaceship will not be able to find any charge anywhere. We also have infinite potential positive and negative voltages but everywhere the voltage is zero.

The dots do not exist at this time. However the equivalent of the universe contain an infinity of plus minus dots which are perfectly superimposed. The pure electromagnetic field of pure nothingness can stay at infinity forever. However, the contraction of the field starts to produce current flows. If the field contacts perfectly symmetrically toward a single point, two dots are produced by the current flows.

The first big bang is the result of the compression of the field. At minimum two dots were produced which separated in space-time. Later big bangs produced more and more dots. For each big bang the number of dots increased and the charge and energy of each dot dropped.

We can also argue that a huge amount of dots were produced by the first big bang. The dots are the product of the current flows. The compression of the pure electro-magnetic field or pure space-time causes currents to flow. The current flows produce mass. The universe we see is the product of the compression and later expansion of space and time.

We can look at space-time as being similar to a static electricity generator. As space is compressed, capacitances increase and huge charges build up. The universe splits into a hysteresis loop with tremendous voltages across the barrier between positive and negative halves of the universe. A point is reached where a current flow occurs. The dot currents puncture the barrier and the big bang is born. The big bang is the initial discharge current between both halves of the universe. The property of mass then comes into being.

After the big bang, the flow of current continues until all the dynamic energy is used up. The universe expands and the impedance of the universe moves toward infinity. The impedance of the primordial universe is:

$$Z_U = \infty \tag{9-14}$$

The impedance of the primordial universe is infinite. There is no current flow and the universe is infinite. The impedance of the universe at big bang is:

$$Z_{U(bb)} = 0 \tag{9-15}$$

In equation 9-15 we see that the impedance of the universe at big bang is zero. This enables the tremendous discharge of current flow between both halves of the universe. Thus:

$$I_{U(bb)} = \infty \tag{9-16}$$

The voltage across the barrier at big bang is finite. We have zero impedance and infinite current. The combination produces a finite voltage across the barrier. The active energy is infinite.

We see that pure nothingness is a complex entity. The universe comes from pure nothingness but pure homogeneous space-time possesses the ability to contract and produce an infinite variety of universes. When a space ship encounters a primordial universe, it cannot find anything nor measure anything. Nothing at all exists. Yet the universe is an electromagnet field which has expanded like a spring. The spring has lost its tension but is has the ability to regain its tension once it starts to compress.

The universe starts to compress and the spring gets stronger and stronger. A point is reached where the spring is full of dynamic energy. Then at full compression, all that energy is released. The electromagnetic field discharges and the universe we know is born.

SECTION 9-2: DOUBLE SLIT EXPERIMENT EXPLANATION

Let us look at the double slit experiment using the Dot theory concepts and GG/MM/Doppler space-time to understand it. Many modern mechanical physicists agreed that the basic element of quantum theory was the double slit experiment. They agreed that it was absolutely impossible to explain the experiment in any classical way, and that quantum mechanics was necessary to explain it. Let us see where this thinking is incorrect.

Young's double slit experiment involved a light source that passed through two slits and hit a wall surface. Light and dark areas were noticed in the light pattern that showed constructive interference and destructive interference within the display.

The slits were made very small compared to the wavelength of the light and the distance to the screen was made large compared to the distance between the slits. This enabled large patterns to be produced which provided a method for indirectly measuring the wavelength of the light.

The Quantum Mechanics two slit experiment provided filters after the light source such that only one photon at a time would enter the two slit apparatus. Evidently the filters were so very cloudy and so thick that most of the photons per unit time were absorbed. Assuming that only single photons emerged from the filter and not dual photon pairs, the photons could only pass through one slit at a time and therefore no interference pattern could emerge.

However, single photons hitting a photographic plate did produce interference patterns. It appeared that the single photon acted like it passed through both slits at the same time and thereby produced interference patterns.

We now have two distinct possibilities. The first solution is a mass/photon oscillation, which causes self-induced interference patterns. This will be discussed in this section. The second solution involves space-time characteristics, which will be discussed in Section 9-3.

In the first case, a single photon will convert into a matter/photon as the speed of the photon drops from light speed to much less. Thus:

$$M_{Xg} = M_O [1-(V/C)^2]^{1/2} \tag{9-17}$$

In equation 9-17, an original photon traveling in space at the speed of light C will be a pure photon of rest mass M_O. It will have gravitational mass in the Y and Z directions perpendicular to the axis of travel. It will only have inertial mass in the X direction. Once we pass the photon through ordinary air, it will gain a slight amount of gravitational mass in the X direction as its speed drops slightly. This is due to increases in the electrical permeability U_O and the electrical permitivity ε_O. When it passes near a star it always has gravitational mass in the Y and Z directions and will bend slightly toward the star.

When we send the photon in a glass filter the light speed will drop dramatically. The index of refraction of Crown Glass is about 1.54 on the average, which makes the speed of light in crown Glass:

$$\text{Crown Glass Velocity} = C/(1.54) = 0.65\ C \tag{9-18}$$

This will cause the gravitational mass in the X direction to become:

$$M_{Xg} = 0.759 M_O \tag{9-19}$$

In equation 9-19 we see that a photon moving through crown glass has gravitational mass in the forward direction of 76% of the rest mass. It becomes a particle wave at that point. The experiment converted an ordinary photon into more of a particle than a photon. Let us now look at the speed of light in diamond.

$$\text{Diamond Velocity} = 0.4C \tag{9-20}$$

This will cause the gravitational mass in the X direction to become:

$$M_{Xg} = 0.9165 M_O \tag{9-21}$$

We see that within diamond the mass of a photon is 92% of its rest mass. Photons traveling through diamond are mostly particles at that time.

We can produce filters that reduce the photon speed quite low. With enough impurities and darkeners added we could reduce the photon speed to nearly zero and capture all the photons. If we reach zero speed, the photon will be a pure particle. In the double slit experiment photons have been converted to particles before they left the apparatus.

The addition of a filter to the stream of photons not only blocks most from reaching the double slit, but also changes them dramatically. Electrons moving at high speeds and then slowing will experience similar effects.

In the first case of the photon, we have produced matter/photons which emerge from the filter with a velocity lower than 0.65C. These matter/photons contain a normal AC photon-oscillating field superimposed upon a spherical matter-oscillating field. The AC field of the matter/photon is a complex dual matter/photon field. In addition, it is a transient changing field. As the matter/photon emerges from the filter at lower than 0.65C, it will build up speed. Gravitational mass will be converted into inertial mass in the X direction. There will be an electrical transient oscillation that changes the matter/photon into a pure photon. This oscillation depends upon the inductance and capacitance of space, the inductance and capacitance of the filter, the electrical effects of the double slits, and the electrical effects of the photographic film used to absorb the photons. Therefore we have a complex transient electrical problem.

If the matter/photon only passes through one slit, the internal conversion oscillation from matter/photon to pure photon will produce self-interference patterns. A single photon or electron moving through a single slit will produce patterns of reinforcement and destruction. The reason being that the photon is an AC field, which changes from a partially spherical shape to a cylindrical shape of a regular photon.

In this section we see that the cause of the double slit problem is due to the conversion from particle/photon to pure photon. In the next section we will look at the space-time effects in the experiment.

SECTION 9-3: DOUBLE SLIT IN SPACE TIME

In section 9-2 we looked at the double slit experiment from a purely electrical circuit perspective. Now, let us look at the double slit experiment again from a space-time perspective. In the electrical analysis of section 9-2, it was felt that a photon produces a self-induced matter/photon oscillation when passing through a slit in the double slit experiment. Yet one would expect the somewhat similar results from a single slit experiment. The double slit does accentuate the interference problem since the photon can pass through one slit while the AC photon field oscillation can pass through both slits. Therefore the explanation in section 9-2 only gives a partial explanation to the problem. Now we can use GG/MM Doppler space-time concepts to better understand the problem.

When we look at things in GG/MM/Doppler space-time, alternative solutions are readily available. We see in the initial space-time study of the Bohr atom, a magnetic repulsion occurs for the electron with itself. In that case we see that a current does flow and this current loop matches the electrical laws for repulsive currents even though only one electron exists.

We also see in the cyclotron that an electron becomes a circular line. However, it is really straight-line segments, which must be focused with a magnetic field to keep it within a circle. The electron exists everywhere in the cyclotron when we approach the speed of light. Even when we don't achieve the speed of light, the Doppler frequency is higher in the front and lower in the rear. The mass or charge exists in the front of the motion. In addition, the Doppler length is far ahead of the electron. We then have a law of physics for fast forwarding of the effects of an object or an electron as it moves in a straight line.

FAST FOWARDING LAW #1: An object increasing speed to velocity V will have an effect in front of it moving with velocity C. The electron accelerating in the cyclotron has an effect in front of it which catches up with it. It will have a mass in front of it and it will have a length in front of it. These will move and form standing waves as they pass themselves. With circular motion at constant velocity, we eventually achieve a steady state waveshape after a few complete cycles. However even at the very beginning, the effect of the electron is felt in front of it at the speed of light.

If we have photons in the double slit apparatus and are sending only one photon at a time through the apparatus, we still have the effect of one particle/wave leaving the apparatus and increasing speed to the speed of light. The one particle wave will pass through one slit. However, the fast forward action will send a wave-front through both slits.

The same is true if we send an electron through the apparatus. As it speeds up it becomes more wave and less particle. Its inertial mass in the X direction moves toward zero as it passes the filter and gains velocity. The same is true within the filter itself. As the electron slows, a certain amount of wave energy is fast-forwarded through the filter. A single electron will pass through only one slit but the wave-front will pass through both slits. This will cause the interference pattern.

Let us return to the electrical characteristics that accentuate the effect. The slits of the apparatus form a capacitor and inductor with the film surface. In general:

$$\underline{C}_A = \varepsilon_0 \, A/L \qquad (9\text{-}22)$$
$$L_A = U_0 \, A/L \qquad (9\text{-}23)$$

In equations 9-22 and 9-23 we see that the apparatus has a capacitance and an inductance associated with the path of the slits. An electron moving through one slit resonates with the capacitance and inductance of the path. After the electron hits the photography film, the space behind it is still oscillating. This will die out as a damped oscillation. Yet as soon as it starts to die out, the fast forwarding effect of the next electron approaches the area. Then the next electron comes.

The presence of an electron in a situation where many are continuously flowing is readily felt when the electron is not present. This is an electrical transient problem. The electron flowing in one slit sees previous electrons flowing through the same slit but more importantly previously electrons flowing through the second slit. Depending upon the speed of flow, a transient electrical effect occurs in addition to the splitting of the wave front.

If an electron just passes a point in space once, we will see a starting blip or magnetic field effect just before it arrives. Then when it arrives, we will see the regular magnetic field effect. Finally after the electron leaves the space, we will see the decaying transient oscillation.

The double slit experiment is a space-time, and an electrical circuit theory problem. The majority of the double slit phenomenon is pure electrical theory with fast forwarding of the effects of the motion of the electron. Let us now look at the case where only photons appear in the double slit apparatus.

The apparatus filters the photon source severely so as to limit the photon flow to one at a time. In order to do this, the filter must bring the light speed of the photons very low and also absorb most of them. When a photon starts to leave the filter and head toward the double slit apparatus, the photon is part photon/part matter. The velocity of the photon is V. The photon has a fast forwarding Doppler effect ahead of it. At the speed of light the photon effect wave-front arrives at the double slit while the photon itself arrives later. The wave-front can pass through both slits while the photon itself only goes through one slit. In addition, the Doppler inertial length extension will fast-forward the photon before it reaches light speed.

We get a fast forwarding effect, transient oscillating effect, some space-time memory effect, and a splitting of the AC electromagnetic wave-front. The more rapid the photon flow, the stronger all of the transient effects will be.

The single photon at a time will see the equivalent of another photon in the opposite slit. We will then get sums and differences in the photographic film.

For the electron in the cyclotron, we get the inductance and capacitance of the loop in which the electron is traveling. In addition we get the space-time memory effect as well, and we get the Doppler inertial length extension in front of the electron. Therefore the four effects cause the fast moving electron to look like a line of charge.

SECTION 9-4: THE ADAPTABLE PHOTON

Let us now study the adaptability of the photon. Let us look at the speed of a photon coming from a planet in our solar system. The photon originated at our sun, bounced off the planet, and traveled directly toward this Earth. Since the photon came from our sun, we can say that the galaxy motion and the sun motion are common mode to everything.

The photon travels toward the Earth with the common mode velocity of C, which is the speed of light in our galaxy/solar system. Throughout the galaxy, the speed of light will vary slightly. Now consider a simple photon that is not much different than one from this Earth.

The photon comes to us with the velocity C. We are moving toward it with the velocity V. The relative speed is:

$$\text{Relative Speed of Photon} = V+C \qquad (9\text{-}24)$$

The relative energy of the photon is:

$$E(\text{photon})\ \text{relative} = h\,(C+V)/\lambda \qquad (9\text{-}25)$$

In equation 9-25 a photon is traveling toward the Earth. The Earth is moving toward it with a velocity V. This cause the photon to be blue shifted due to the relative velocity of the Earth and the sending planet. We can assume that the sending planet is stationary for this example. Otherwise we can give both a velocity and V is the relative velocity of the two planets.

We see that the Earth moving toward the photon causes it to be blue-shifted, and it gains energy. If the Earth were moving away from the photon, it would be red-shifted and lose energy.

The photon is adaptable. When it reaches our measuring instrument, it will be moving at C relative to the Earth. It has higher energy but lower relative speed then when it was further away. What happened?

Far from the Earth, the photon was moving at the speed C and the Earth was moving toward it at the velocity V. The photon is composed of plus and minus dots. These dots tend to form a plane surface in which the dots oscillate from the circumference toward the center. If you look at it sideways, it will look like a flat surface which has a tail as it oscillates.

The photon is very happy and moving at the speed of light C for our solar system. The Earth meanwhile is moving toward the photon. The Earth's gravitational field contains free space dots in front of it. These are compressed in the direction of motion and elongated opposite to the direction of motion.

The density of the free Earth dots is high close to the surface of the Earth and decreases as we move further away. Far in outer space, the density of free dots is low. Our happy photon starts to encounter higher concentrations of the Earth dots. We start to get interactions far out in space. Our happy photon starts to get compressed by the Earth's gravitational field. It absorbs free dots and starts to slow relative to the Earth's gravitational field.

Time passes. The happy photon absorbs some of the Earth's free dots. It gains energy but slows relative to the Earth. Finally after a long time, the photon reaches our measuring instrument. The speed is measured as C. The energy is measured as blue shifted. The net result is that the photon adapted. It lost relative velocity and gained relative energy.

One alternative to the Michelson/Morley experiment is that the photon adapted. It adapted in the direction of motion and it adapted in the perpendicular direction. The reason that the speed of light appears constant for all reference frames is that the photon adapts to all the reference frames. Before the photon reached the Earth, it adapted to the Earth's gravitational field and free space dots.

Now it can be argued that we do get some adaptability and some shrinkage of distance. The photon adapts, but length and time also adapts. The net result is a slightly different set of space-time equations in which photon adaptability has to be added. Yet this may only be a secondary effect to the main effect of space/time variability. Of course concentrations of free space dots do change the speed of light. The more free space dots, the lower will be the light speed in the particular area.

We live in a strange universe where references are hard to find. Fortunately most things upon our Earth move slowly compared to the speed of light. The gravitational field adapts to the motion of the Earth, the Sun, and the Galaxy. There are many distortions that we cannot see and measure. Most things are common mode and we exist in a world where only differences count.

SECTION 9-5: SPACE TIME QUANTUM MECHANICS

Let us look briefly at Quantum Mechanics and see how it relates to the Dot theory with GG/MM/Doppler space-time. There are five quantum numbers associated with the electron in orbit. Let us make a table of the quantum numbers.

QUANTUM NUMBER	DEFINITION
n	Energy- size of elliptical orbit
l	Orbital Angular Momentum
M_l	Orbital Angular Momentum along Z axis
s	Spin Angular Momentum
M_s	Spin Angular Momentum along Z axis

The quantum number (n) is associated with the particular Bohr orbit. The orbital angular momentum is associated with the linear center of gravity Doppler shift, and the spin angular momentum is associated with the Inertial Center of gravity Doppler shift.

According to the analysis in Doppler space-time, there is a Doppler center of gravity shift associated with a moving mass. The center of gravity of a perfect sphere in motion will be shifted in the direction of motion. The electron contains both a mass and a charge. The motion of the electron in orbit around the proton causes a magnetic field along the Z-axis. Now we have both a mechanical and an electrical problem. The Doppler-shift in the mass causes a space-time shift in the mechanical angular momentum and the magnetic angular momentum.

There is a balance of electrical and mechanical forces within the Bohr orbit. One Doppler shift produces a slightly different effect than another Doppler shift. In quantum mechanics this is called space quantization. However, it is merely a simple electrical verses mechanical balance of force problem.

Let us now look at the spin terms. As the electron spins, there is a Doppler Center of Inertial mass shift. As shown in the discussion of rotational motion, the Doppler Center of Inertial mass shift is to the outside of the stationary Inertial Center of Gravity, ICG in some cases, inside the ICG in other cases, and at the same exact location is still other cases. This depends upon the centripetal and centrifugal forces, the internal binding energy and photon energy forces, and the external applied forces. In general in pure space for massive systems, the centrifugal forces will be stronger and the Doppler will be shifted outward, and Galaxies will continue to expand.

The space-time shift of the inertial Center of Gravity causes the spin angular momentum number and the corresponding magnetic quantum number.

The entire system of quantum numbers tends to interlock with each other. A planet will not orbit a sun unless a balance occurs between the Doppler center of mass shift and the Doppler inertial center of gravity shift. The planet will cycle toward the sun or fly into space. The same is true of the atoms.

As the electron moves toward the proton the perpendicular space-time mass will increase with velocity while the forward mass will decrease. The moment of inertia will also increase. The same principles occur in planetary motion or within the Bohr orbit. The only thing on the small scale is that the electrical force produces similar results.

It should be noted that the four quantum numbers are not sufficient. The electron will experience summer and winter. It will oscillate in the Z axis direction, just as the Earth does. This will account for the fine spectra.

In general, if you know how to solve the GG/MM/Doppler space-time equations for this Earth, then you can apply the results to the Bohr Orbit. Of course you must take into account the dual nature of the mechanical and electrical forces. The Bohr Orbit works the same way as the Earth but the equations are somewhat different.

If you look at the huge Earth size protons in outer space with the huge Earth size electrons, you will find the hydrogen atom full size. More often, you will find huge neutrons with strong inner spins.

SECTION 9-6: BOHR ORBIT IN SPACE TIME

Let us look at the Bohr Orbit again using space-time concepts. The inability of scientists and mathematicians to improve upon the Bohr Orbit lead to the field of Quantum mechanics. In this field highly complex mathematical concepts were used to explain how the atom worked. Let us now look at the Bohr orbit in space-time. In the case of the Bohr Orbit, the basic law was:

$$MV^2/R = KQ^2/R^2 \tag{9-26}$$

This equation stated that the coulomb attraction between the proton and the electron was equal to the centrifugal force of the mass of the electron in orbit.

We could add the motion of the proton to it and increase the complexity but the basic equation is incomplete.

The gravitational attraction between proton and electron was left out. This is a small term compared to the electrical force but it is a term none-the-less. However, the big term is the electrical term. An electron in orbit produces a current loop. Currents flowing in the same direction in electrical wires cause attractive forces whereas currents flowing in opposite directions cause repulsive forces. Thus:

$$F = U_O A_Q I_A I_B \tag{9-27}$$

The attractive or repulsive force between two parallel current carrying wires is equal to the electrical permeability U_O times the current, I_A of one wire times the current, I_B of the second wire times a constant A_Q which depends upon the geometry of the wires. If we look at the electron in the Bohr orbit, the distance traveled by the electron is $2\pi R_{Bohr}$. For the moment let us assume the circular loop constant is A_Q Thus:

$$GM_P M_E/R^2 + K Q Q/R^2 = (MV^2)/R + U_O A_Q (I_E)^2 \tag{9-28}$$

Where the electron current flow I_E is:

$$I_E = QV/(2\pi R) \tag{9-29}$$

Equation 9-28 is a more complete Bohr Orbit equation and shows the repulsive force $U_O A_Q (I_E)^2$. Notice that it is both a mechanical and an electrical equation. The actual equation or set of equations will be much more complex due to the effects of the proton and the electron spins, etc. A series of equations can then be written to produce exact solutions for the Bohr orbit. Ordinary classical physics with space-time Doppler shifts can be used to calculate the complete Bohr atom on a computer program.

Over the years some physicist's thought that the electron would head right into the proton except for fancy quantum physics rules and regulations. They did not account for the electrical current loop term.

As the radius R gets smaller and smaller and the velocity get larger and larger, the current loop become a lot stronger in repulsion than the mechanical centrifugal forces. The electrical loop will tighten in a space-time manner and this will produce heavy space-time repulsive forces.

The electrical space-time forces are so much stronger than the mechanical space-time forces. The advantages of the space-time Bohr atom, which would require a separate book, are that everything can be calculated. There is nothing fancy that requires anything above Engineering school.

The electron has a Doppler mass shift in the main orbit and an electrical charge Doppler shift as well. When we look at the spins, we find a phase angle Doppler shift of the center of Inertia of the mass and also a Doppler shift of the center of Inertia of the charge. As we saw in the gyroscope example, the various accelerations cause momentum unbalances that cause the objects to precess at right angles to the plane of rotation. Therefore the electron has the same effects. Of course the protons have similar effects as well. Compete equations can be written about the Bohr orbit and a computer can produce exact answers using simple classical techniques.

If we look at the lowest state of the hydrogen atom we see that two possibilities exist. If one electron orbits in one plane, then another electron can orbit perpendicular to that plane.

In order to maintain a balance of electrical forces, the spins of each loop must be different. In this manner they can remain perpendicular. Therefore the two different states are really an electrical problem of balancing two current loops so that they don't become one loop or destroy each other. Therefore the spins must be opposite to prevent the loops from coming together.

If we go to higher orbits, we notice that mechanical equations produce equal energy for various orbits. However, the electrical loops do not obey the same mechanical laws. The energy of various elliptical orbits will not be exactly the same.

The purpose of this section is not to solve the Bohr orbit exactly but to point out how to do it and to point out the mistakes of the past. The Dot theory, with GG/MM Doppler space-time and the correct equations enables classical techniques to be used by anyone to accurately understand the various atoms.

The electron spin has two values for the Bohr orbits. As we look at the electron in the various orbits, we have only two non-radiating choices. The electron will act like the moon. It will lock with its face toward the Orbit. This only gives it two choices. If the electron spun on its axis faster than a lock step, it would slow and radiate.

The problem with quantum mechanics is that it does not give the reason for the spin only having two values. Now that we understand that the moon is lock stepped, the electron in the Bohr orbit can exhibit the same tendencies.

The moon long ago most likely had some spin on it. However, it radiated that energy and now only shows one face toward us. Likewise some day this Earth will be locked-stepped toward the sun and only show one face like Mercury.

Let us calculate the constant A_Q for the circular loop in the Bohr orbit. Let us assume this constant is the same value as for the electron in the neutron orbit. When we move at the speed of light, the electrical coulomb attractive forces will be equal to the electrical magnetic forces. Thus:

$$K Q^2 / R^2 = U_O A_Q (I_E)^2 \qquad (9\text{-}30)$$

$$K Q^2 / R^2 = U_O A_Q Q^2 V^2 / (2 \pi R)^2 \qquad (9\text{-}31)$$

Since $V = C$, and $K = 1/(4\pi \varepsilon_O)$, and $U_O \varepsilon_O = 1/C^2$, we get:

$$A_Q = \pi \qquad (9\text{-}32)$$

Thus the Bohr Orbit equation is:

$$GM_P M_E / R^2 + KQ^2/R^2 = (MV^2)/R + U_O \pi [QV/2\pi R]^2 \qquad (9\text{-}33)$$

Notice when we reach the velocity of light, the electrical terms cancel out. We then are left with only the mechanical terms of planetary motion on a small atomic scale.

SECTION 9-7: ELECTRIC CHARGE IN SPACE TIME

In this section let us look at a space-time viewpoint of a dot. Let us use the Michelson/Morley test results and relate them to the electric field in space-time. The Dot theory has defined the dot as a charge Q_D that extends to the radius of the universe and appears as a charge Q at R_U. Locally, it is an electro-photon which has a charge of 1.4071E-60 coulombs in the here and now and 6.60218E-19 Coulombs at R_U. It has an equivalent rest mass of 1.3717E-72kg in the here and now. The dot is always moving outward at the speed of light at the radius of the universe.

In Chapter 7 section 7-3, we learned that the inertial length of an object moving to the right, or front is:

$$L_R = L_{Xg} [\, C/(C-V) \,] \tag{9-34}$$

From equation 9-34 we see that the inertial length to the right gets very large as compared to the gravitational length. The length of the object to the left or rear is:

$$L_L = L_{Xg} [\, C/(C+V) \,] \tag{9-35}$$

From Equation 9-35 we see that the inertial length to the left only becomes half the gravitational length. The ratio of the gravitational length in the X direction to that in the Y direction is:

$$L_{Xg}/L_{Yg} = [1-(V/C)^2]^{1/2} \tag{9-36}$$

In equation 9-36 we see that the length in the X direction shrinks as compared to the Y direction as the velocity increases. For GG/MM/Doppler space time, the Y and Z dimensions are invariant until near the light speed barrier. Therefore:

$$L_{Xg} = L_O [1-(V/C)^2]^{1/2} \tag{9-37}$$

By combining Equations 9-37 & 9-38, we get:

$$L_{XR} = [L_O[1-(V/C)^2]^{1/2}] \cdot [C/(C-V)] \tag{9-38}$$

We see that as we move toward the speed of light, the inertial length to the right in the X direction reaches toward infinity. It reaches the radius of the universe.

By combining Equations 9-37 and 9-35 we get:

$$L_{XL} = [L_O[1-(V/C)^2]^{1/2}] \cdot [C/(C+V)] \tag{9-39}$$

We see that as we move toward the speed of light, the inertial length to the left in the X direction reaches toward zero.

Let us now write the equations of an object which moves spherically at close to the speed of light at the radius of the universe and which has a tiny size locally.

$$L_{Xg} = L_{Yg} = L_{Zg} = L_O [1-(V/C)^2]^{1/2} \tag{9-40}$$

Equation 9-40 states that an object moving spherically at the velocity V in all three directions simultaneously experience shrinkage in length according to Einstein's orbital shrinkage formula. By using Equation 9-40 we have converted linear motion and linear space-time into spherical motion and spherical space-time. We see that the object becomes a dot when it is moving outward at the speed of light in all directions simultaneously. We see that the dot is really the Doppler inverted image of a charged light sphere.

Let us now look at the Doppler spherical equivalent of the dot itself. The length to the left would be the equivalent of the size of the dot. The length to the right would be the size of the universe. We can now write equations for the dot dimensions in terms of the root mean square value of the Doppler dot. Thus:

$$R_{D(RMS)} = (R_D R_U)^{1/2} \qquad (9\text{-}41)$$

In Equation 9-41, the root mean square (RMS) value of the dot radius is the geometric mean of the local dot radius and the radius of the universe. We can now write the space time equation for the radius of the universe in terms of the RMS radius of the dot. Thus:

$$R_U = R_{D(RMS)} [1-(V/C)^2]^{1/2} \cdot [C/(C-V)] \qquad (9\text{-}42)$$

Equation 9-42 is an increasing function of velocity. At V= 0.9C, R_U has a value of $4.359 R_{D(RMS)}$, while at V=0.9999C, R_U has a value of $141.42 R_{D(RMS)}$. In order to achieve a large ratio of R_U to $R_{D(RMS)}$, the dot must travel at almost light speed C.

We can now calculate the dot radius in terms of the dot RMS radius. Thus:

$$R_D = R_{D(RMS)} [1-(V/C)^2]^{1/2} \cdot [C/(C+V)] \qquad (9\text{-}43)$$

Equations 9-42 & 9-43 enables us to understand why the dot is traveling at the speed of light outward and why it is so tiny and so large at the same time. In equation 9-42 we see that the dot reaches to the radius of the universe. In equation 9-43 we find that the dot size is basically zero.

From equation 9-42 we could calculate the exact light speed required to achieve R_U for a certain size dot. We do not have enough information to calculate the dot size directly. However, by comparing the Doppler dot equations with the Doppler charge equations we will be able to do this.

Both the plus dot and the minus dot are discharging into each other. They are both moving at the speed of light C but this movement is far from the center of the dot. Therefore the dot itself can be stationary or a variable speed. The dot current flows and the dot outer shell moves at the speed of light but the center of the dot can stand still. The dot is simultaneous a huge thing and a tiny thing.

We see that the dot size depends upon the radius of the universe and the velocity of the dot light sphere that is moving at approximately the speed of light. This velocity will only be slightly different from light speed. Let us now write the Doppler equation for the dot charge. This will enable the dot size to be calculated as well. The dot charge equation is:

$$Q_{D(Ru)} = Q = [Q_0 [1-(V/C)^2]^{1/2}] \cdot [C/(C-V)] \qquad (9\text{-}44)$$

In equation 9-44 we see that when using the standard Doppler equations, a term Q_0 appears, which is the actual charge of a dot in terms of the midpoint or RMS value. This stands as the geometric mean between the little dot charge at the dot point and the larger dot charge at the radius of the universe.

Let us now write the equation for the dot at the present location in the here and now.

$$Q_D = [Q_0 [1-(V/C)^2]^{1/2}] \cdot [C/(C+V)] \qquad (9\text{-}45)$$

In equation 9-45 we have the dot charge in the here and now. We now have two equations with two unknowns. The charge Q_0 and the exact velocity of the charge sphere bubble. We do know Q_D from previous calculations and we do know Q. By dividing equation 9-45 by 9-44 we get:

$$Q_D / Q = (C-V) / (C+V) \tag{9-46}$$

Using $Q_D = 1.42186\text{E}-60$ and $Q = 1.60218\text{E}-19$, we get:

$$[C-V] / 2C = 8.81836\text{E}-42 \tag{9-47}$$

$$C-V = C (17.63672\text{E}-42) \tag{9-48}$$

$$\Delta V = 5.28735\text{E}-33 \tag{9-49}$$

In the same manner we can write the dot radius equations:

$$R_U = [R_O [1-(V/C)^2]^{1/2}] \cdot [C/(C-V)] \tag{9-50}$$

$$R_D = [R_O [1-(V/C)^2]^{1/2}] \cdot [C/(C+V)] \tag{9-51}$$

In Equations 9-50 and 9-51 the inner and the outer dot dimensions are shown. We have not calculated R_O. It is the same R_O as with the dot charge. We now need some fancy approximations since the hand calculators cannot handle such numbers. You need computer programs with 50 place calculations. However an approximation can be found.

The simple solution is to compare the distance equations to the charge equations. We find:

$$Q_D / Q = R_D / R_U \tag{9-52}$$

Equation 9-52 enables us to calculate the dot size.

$$R_D = [Q_D / Q] \cdot R_U \tag{9-53}$$

$$R_D = 1.31306\text{E}-15 \tag{9-54}$$

In Equation 9-54 we see that since $R_U = 1.40901\text{E}26$, the dot radius comes to approximately the size of the neutron. It was assumed throughout this book that the size of the dot is very tiny. Now we find that the dot size is a window through which dot current flows. The current flows to a point or dot but there is a degree of uncertainty of where the exact point of flow within the dot is. There are huge amounts of dots merged together within R_D. We can still think of the dot as a pure dot. However, from the Doppler analysis we see that the dot is an opening in space-time. It is a node. It is an electrical junction point in space-time.

Let us now calculate Q_O as the root mean square of the dot charge, Q_D and the charge Q. Thus:

$$Q_O = [Q_D \cdot Q]^{1/2} \tag{9-55}$$

$$Q_O = 4.772919\text{E}-40 \tag{9-56}$$

In the conversion from mass to charge we found that:

$$M = QC \tag{9-57}$$

At that time we did not know what charge related to what mass. The dot charge was too small and the charge Q was too large. However, the charge Q_O is just right. Thus the mass of the electron is:

$$M_E = 9.10939\text{E}-31 \tag{9-58}$$

We can now match the mass of the electron as:

$$2\pi Q_O C = 8.99048\text{E-}31 \tag{9-59}$$

Thus to within an error of 1.29%:

$$M_E = 2\pi Q_O C \tag{9-60}$$

We see that the electron is the standard mass to charge conversion formula to within an error of approximately 1.3-percent. We notice that we got the same 1.3-percent error in the calculations of Chapter 5 section 1. There is a difference between the here and now calculations of standard electrical theory and the dot theory calculations using the radius of the universe.

From an Engineering viewpoint, the error term is acceptable. The length R_O will also be:

$$R_O = [R_D R_U]^{1/2} \tag{9-61}$$

$$R_O = 4.42172\text{E}5 \tag{9-62}$$

The distance R_O is the geometric mean distance between the dot size and the radius of the universe. Let us now look at the value of the geometric mean mass for the universe.

$$M_O = [M_D M_U]^{1/2} \tag{9-63}$$

$$M_O = 5.2454\text{E-}10 \tag{9-64}$$

The charge Q_O, the distance R_O, and the mass M_O are the root mean square values for charge, distance, and mass respectively. The significance of these numbers necessitate future study.

SECTION 9-8: FASTER THAN LIGHT

The speed of light in the Michelson/Morley experiment is the speed of photon travel. Photons are configurations of dot energy, which move at the average speed of measured light of 2.99792E8 meters per second in vacuum. The dots themselves have a slight front to back motion so that the dot speed is somewhat larger than the photon speed.

The speed capability of the universe is infinite. We can have coexisting universes of extremely high light speed within our space-time bubble. This is for a multi-light-speed solution. The single light speed solution permits the universe to change from an infinite light speed universe to a constant light speed universe.

The constant light-speed solution to the universe still has infinite light speed capabilities. These capabilities permit simultaneous reactions all over the universe. An event such as the big bang would be felt everywhere simultaneously.

These reactions are not carried by individual photons. They are perturbations of the entire electro-magnetic field of the universe. An explosion of a star in one galaxy can disturb the entire space ether, which is the total electro-magnetic field. This causes a disturbance all over the universe. In effect, the entire universe suffers an earthquake of sorts.

The geometry of groups of dots limits the photon speed to the speed of light that we measure. We cannot cause the photon to readily move faster than its normal speed. The individual dots themselves are not bound to the same rules and regulations as photons. The dots of the primordial universe formed photons of extremely high light speed. Therefore it is quite possible for the dots themselves to move faster than light.

We now have two ways of exceeding the speed of light. A perturbation of the entire space ether will travel up to nearly infinite speed. In addition, the dots themselves tend to move at our light speed but can flow up to infinity light speed. This would produce very high-speed dot electromagnetic fields, which could transmit energy and intelligence all over the universe at up to infinite light speed.

This higher than light speed capability converts the entire universe into a singular intelligence field or space-time mind. Of course the entire spectrum of dot energy permits dots of decreasing charge from what we have. Our dots will form photons at our light speed. Let us look at a chart of dots of higher light speed.

Charge	Velocity	Mass	Energy
Q_D	C	M_D	$M_D C^2$
$0.5 Q_D$	$2C$	$0.25 M_D$	$M_D C^2$
$0.25 Q_D$	$4C$	$0.0625 M_D$	$M_D C^2$
$0.125 Q_D$	$8C$	$0.0156 M_D$	$M_D C^2$

The above chart illustrates the dot charge, photon velocity, dot mass, and dot energy for a spectrum of dots within a multi-light-speed universe. In this solution, the energy spectrum at all light speeds are constant. A light speed $2C$ universe would have the same energy as our universe. The total energy of the universe would be infinite. We could also chart a declining energy distribution such that the energy at light speed $2C$ would be half our energy. Then the total energy of the entire universe would be twice our energy.

For the single light speed solution, the above chart shows us dot charge, photon velocity, and dot mass for the same energy. When we existed at twice our light speed, our charges were lower. This is for the same number of dots in the universe. The chart provides us with an alternative for the doubling of dots per universe. The dots could have remained the same but the light speed dropped and the charge per dot increased.

The important thing is that our dots tend to produce photons, which operate at our light speed. The dots themselves can lose charge and move faster. Likewise, many other dots can coexist with us. We can also have stars and planets of higher and higher light speeds. At the highest light speeds we obtain standing waves of energy and intelligence which encompass the entire universe.

We could exist within the higher light speed energy. In addition the higher light speed energy could be self-contained within us. Alternately the highest light speed energy could extend to infinity.

This book is primarily devoted to a singular light speed universe with higher than light speed capabilities. Our photons are limited to our light speed. They can be looked on as little airplanes of dot energy. The dots are more capable than the photons although it would seem that the light speed C dot matches the light speed C universe and higher light speed dots form patterns within higher light speed universes.

Recent experiments have produced results, which indicate higher than light speed capabilities. Physicists have experimented with laser pulses through Cesium vapor, which produces reaction speeds of over 300 times the speed of light.

This book is ready for printing but lately there have been various physicists experimenting with higher than light speed effects. Therefore this paragraph has been added to bring an understanding of the initial findings to the reader.

The Fast Forwarding Law will bring the image of the laser pulse to the Cesium vapor long before the pulse has actually arrived. Therefore the effect occurs in a weakened state long before the cause arrives. This is because of the Doppler image and the Doppler forward length.

In addition, the Cesium vapor will still contain energy and oscillations from the prior pulses. The Fast Forwarding image of the new pulses will continue the oscillations from the prior image. Therefore the solution is the same or similar to the Double slit experiment.

Since the observed phenomenon is a weak pulse as compared to the actual pulse, it can only be concluded that the reason for the faster than light effects is the Doppler forward image of the forming laser pulse.

It is unlikely that higher light speed dots were active during the experiments. Yet, this is always a possibility. If the higher light speed dots were active, then the observed image would be weaker.

More unlikely is a whole body disturbance of the space ether. The explosion of a star will produce fantastic disturbances, which propagate throughout the universe at ultra high light speed. Some of the disturbance will be carried by the nearly infinite light speed dots existing in free space.

The laboratory experiment most likely will not cause enough local disturbance to be felt in the laboratory. However, it is always possible that a very high intensity disturbance can shake up the space ether enough to be felt nearby.

In conclusion, it appears that the effect was caused by the Fast Forward Doppler image. However, our universe does have infinite light speed capability.

CHAPTER 10: THE FM DOPPLER MODEL UNIVERSE

SECTION 10-0: INTRODUCTION

In this Chapter we will look at a different model of the universe from a standard frequency modulated (FM) perspective. This will help us to understand the universe we live in.

SECTION 10-1: FM DOPPLER SPACE TIME

Let us look for a purely electrical solution to describe the workings of the universe. In Chapter 8 we found that the Einsteinian clock solution was good for orbital motion. The GG/MM Doppler space-time solution was also shown to match the Einsteinian solution for orbital motion. Therefore Einstein was correct in his orbital clock solution.

Einstein's mass equation is correct as well for steady state orbital type motion such as the cyclotron. However, in outer space electrons and positrons are produced all the time by the collisions of strong photons with energy levels of the order of the electrons themselves. These are transient conditions and the electrons and positrons will achieve close to the speed of light without any great amounts of added energy.

It is obvious that the equations for the energy of an electron which travels almost at the speed of light in pure free space is:

$$E_{E@C} = M_0C^2 + \tfrac{1}{2} M_0C^2 + \tfrac{1}{2} M_0C^2 \qquad (10\text{-}1)$$

The first term is the rest energy, the second the kinetic energy, and the third is the space-time energy, which can be considered as increased mass. The increased mass can be considered inner rotational energy where the plus dots rotate one way and the minus dots rotate the other way thereby shrinking the electron to a line. The above solution will give us the proper answer for photon to mass conversions in atomic physics as well as for classical physics depending upon rotations, gravitational fields, etc. The energy is between $3/2\, M_0C^2$ and $2M_0C^2$ as we near light speed for the GG/MM/Doppler solution.

Let us now search for an electrical solution that is the equivalent of the mechanical solution. Since the universe can be described electrically or mechanically, both solutions are acceptable. We can move into the electrical world and look at the problem of the variation of clock frequency, and mass with velocity.

The solution to the problem is a frequency modulated Doppler solution. We can look at a single proton oscillating at a point above an Earth which is moving slowly around a sun and whose galaxy is moving slowly in the universe. We can assume that the proton is stationary.

The proton itself produces a perfect clock. The atomic clock, the M/M apparatus, and the Cesium clock are not perfect clocks. You will get a lot of imperfections in any of the usual clocks. With the proton, you have the best. Of course you cannot measure the proton frequency readily but we know its frequency. The proton wavelength is:

$$\lambda_P = h / (M_P\, C) \qquad (10\text{-}2)$$

In equation 10-2 we have an oscillating proton with a fixed wavelength determined by Plank's constant (h), the mass of the proton (M_P), and the speed of light [C]. A proton in free space and standing still will oscillate at this frequency. The proton stands as a perfect frequency source. If we move the proton, we will see gravitational waves coming from it. Thus:

$$\lambda_{P@V} = h/ M_P V \qquad (10\text{-}3)$$

Usually electrons are used for deBroglie wavelength experiments. However, the proton is our reference source for this analysis. The proton frequency is:

$$f_{O(P)} = [M_P C^2]/ h \qquad (10\text{-}4)$$

The deBroglie frequency is:

$$f_{deB} = [M_P C V]/ h \qquad (10\text{-}5)$$

We find that the deBroglie frequency for the proton is a Doppler shift. Thus:

$$f_{deB} = f_{O(P)} [V/C] \qquad (10\text{-}6)$$

The deBroglie frequency of the proton is the Doppler frequency variation as it moves. The same is true for the electron in motion. Both the proton and the electron in motion are experiencing a FM Doppler shift.

The proton in motion in the forward direction has a larger frequency than that in the rearward direction. We see the forward Doppler frequency shift. We usually see it with the electron. As the electron comes toward equipment with reasonably stationary electrons within it, the equipment picks up the deBroglie wavelength and not the basic frequency. The basic frequency is common mode to all things. The basic proton frequency is common mode to everything. We do not readily see it. We only see the Doppler frequency shift of the proton or electron.

It then becomes quite obvious that the proton is a frequency source. It is oscillating as a perfect sinusoidal when absolutely stationary. As we move the proton, we start to get FM modulation and Doppler shifts.

Standard FM modulation electrical analysis applies to the proton. We can then look at electrical theory to find out how the proton varies in mass and energy with velocity. Since the proton acts perfectly like pure electrical theory, we have known solutions. FM Doppler space-time will provide us with an exact answer. This makes it superior to GG/M/M Doppler space-time in which we had to figure out what the various solutions would be. In FM Doppler space-time, textbooks have already been written with most of the solutions. It is also easy to create electrical experiments in the lab to produce other solutions.

Let us now look at the problem of the clock in motion from a FM Doppler viewpoint. We can use our perfect proton clock, which is the best that can be achieved. We can run the proton clock in orbit and also in pure space.

If we look at the proton as a pure frequency source, we see that when it is standing still, it produces a singular frequency. As we move the proton to the right, it produces a Doppler to the right. At the same time we get another Doppler to the left. The FM equivalent would be:

$$f_{Right} = f_O [1+(V/C)] \qquad (10\text{-}7)$$
$$f_{Left} = f_O [1-(V/C)] \qquad (10\text{-}8)$$
$$f_{RMS} = f_O [1-(V/C)^2]^{1/2} \qquad (10\text{-}9)$$
$$f_{Arithmetic\ Mean} = f_O \qquad (10\text{-}10)$$

We see that the moving proton has a FM Doppler root mean square equivalent, which is the same as Einstein's equations. We also see that the arithmetic mean of the oscillating proton is constant. This is true of all systems where the velocity is low. The FM Doppler model gives slightly different results than the GG/MM/Doppler answers. However, there are enough similarities to match them pretty well.

Until we reach a velocity above 0.9C we can consider that the proton is invariant with respect to time. We can then look at the universe from the point of view of an invariant time clock.

In this solution we are not dealing with the M/M apparatus but with pure electrical theory. Pure electrical theory states that a pure time clock is invariant with velocity. The arithmetic mean frequency of the proton is unchanging. The proton itself is then a perfect clock. We can look at the universe from this perspective and then see what lies beyond this as the light speed of the universe increases to C.

If we plot the Doppler frequencies to the left and the right, we get a triangle. Thus:

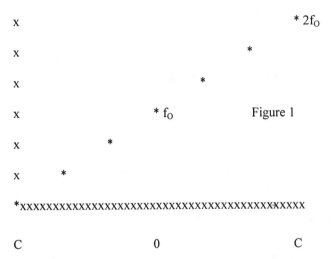

Figure 1

In figure 1 we see that when the velocity is zero, we get a frequency of f_0. However when the velocity is C we get a frequency of zero to the left and $2f_0$ to the right.

This is a standard Frequency modulated electrical system. Thus:

$$\text{Sin}(f_0.t) \, \text{Sin}[f_0.(V/C)t] = 0.5 \, \text{Cos} \{(f_0)[1-(V/C)]t\} \, . \, 0.5 \, \text{Cos}\{(f_0)[1+(V/C)]t\} \quad (10\text{-}11)$$

We see that the GG/MM/Doppler space-time solution is reasonably identical with the electrical FM solution. The Doppler shift is causing the original operating point at f_0 to shift to the right and the left. The arithmetic mean will be f_0 while the geometric mean will be as per Einstein's equation.

The only thing missing from the solution is the energy change from zero velocity to the speed of light. For a true solution, the energy must become twice the amount.

The world of FM electrical theory has no amplitude changes. This means that you cannot change the energy of the system by changing the amplitude. A one hundred hertz sinewave of 120 volts produces the same heat and energy as that of a 60-hertz sine wave of 120 volts rms. The peak values are:

$$V_{Peak} = (1.414)V_{RMS} = 169.7 \text{ Volts} \tag{10-12}$$

The power across a one hundred ohm resistor for a 169.7 volt peak value sinewave is:

$$\text{Power} = V_P^2 / 2R = 144 \text{ watts.} \tag{10-13}$$

The energy is the power times the time. Thus:

$$\text{Energy} = [V_P^2 / 2R]t \tag{10-14}$$

The only way to double the energy in the FM Doppler system without changing the amplitude is to change the waveshape from a sinusoid to a square wave. Thus:

$$\text{Power (square wave)} = V_P^2/R = 288 \text{ watts} \tag{10-15}$$

$$\text{Energy} = [V_P^2 / R]t \tag{10-16}$$

We see that the solution to the problem is a FM Doppler where the wave-shape changes from a sinusoidal at rest to a square wave at light-speed C. This produces twice the energy and is the correct solution to the problem. Now we can understand what happens as we move a mass faster and faster. We produce a spectrum of frequencies due to the conversion of the wave-shape from a sine wave into a square wave. We start with the frequency f_O. This will be modulated both higher and lower. Then the conversion from sine wave to square wave will produce a spectrum of frequencies. At the speed of light we have an infinite series of frequencies to the right.

At the beginning of this book, the Author stated that the solution to space-time was nonlinear. Now we see that we have an infinite series solution to the problem. We see that the universe operates as a pure FM Doppler universe until we get close to the speed of light everywhere. When this occurs, the FM Doppler frequency of the universe will get faster.

The increasing frequency does not produce additional energy. The amplitude must start to change as soon as the universe reaches about ninety percent of the speed of light. This non-linearity of the FM Doppler solution prevents further compression of the universe. It will then explode. The velocity at big bang is:

$$\text{Velocity at Big Bang} = C \tag{10-17}$$

In equation 10-17 we see that we have achieved almost the speed of light when the mass of the big bang is twice the mass at little bang. The present day mass of the universe tends to be close to that at little bang.

We see that the FM Doppler equations and the normal classical physics equations permit us to calculate the velocity at big bang.

SECTION 10-2: MASS OF UNIVERSE AT BIG BANG

In Section 10-1, it was concluded that in order to accelerate an object close to the speed of light, twice the energy as the rest mass was necessary. Thus:

$$M_{(V=C)} C^2 = 2 M_O C^2 \tag{10-18}$$

In equation 10-18, an object is brought close to the speed of light by adding photon energy equal to its rest mass. When we look at the universe at the big bang, we find that everything was moving near the speed of light at that time. Therefore the total energy at big bang was:

$$E_{BB} = 2 E_O \qquad (10\text{-}19)$$

Since the energy at big bang is twice the energy as now. The same is true of the mass:

$$M_{BB} = 2 M_O \qquad (10\text{-}20)$$

The FM model electrical solution to the universe was shown in Section 10-1. This is the FM Doppler solution. In this solution a stationary object can be represented by a pure sine wave. A proton in space, which is not moving, is a sine wave in the electrical universe.

An object that is moving becomes a FM modulated sine wave. The amplitude of the moving sine wave remains constant. However, there is a Doppler frequency to the left and to the right. These become FM side-band frequencies for the FM Doppler model. The center of mass of the object shifts to the right where the higher frequency sine waves are. As explained in section 10-1, as the velocity of the object increases, the sine waves change slowly into square waves. This increases the mass/energy without changing the amplitude.

When we look at photons, we no longer have pure sine waves. To the left of the photon is nothing. To the right of the photon is a square wave. As objects move faster and faster they go from sine waves to square waves. The difference in energy between a square wave and a sine wave is:

$$E_{\text{Square Wave}} = 2 E_{\text{Sine Wave}} \qquad (10\text{-}21)$$

In equation 10-21, a mass represented by a square wave, has twice the energy as one represented by a sine wave. The square wave has an infinite series of harmonics. An ordinary photon has a wavelength and a fundamental frequency but it also contains an entire spectrum of higher and higher frequencies.

When we compress the universe to minimum radius at big bang, all the sine waveshapes have been crushed into square waves. The frequency modulated Doppler universe contracts at constant amplitudes for photons, neutrons, etc., while increasing the frequency toward infinity.

The increase in frequency for the same wave-shape at constant amplitude does not change the energy level of an electrical circuit. The higher the frequency, the more energy the photon. When we look at the universe as a whole, the crushing of the waves is common mode.

In the Dot theory as we crush the universe to a pinpoint, the energy increases toward infinity. This is a common mode effect. However, as far as differences are concerned, the only thing important is that the mass at big bang is twice the mass at little bang or full expansion. The common mode change in energy of the universe does not effect the gravitational calculations. Only the here and now mass is important.

The mass of the universe when heading toward the big bang is increased at constant amplitude until the sine waves become square waves. The mass of the square wave is:

$$M_{\text{Square Wave}} = 2\, M_{\text{Sine Wave}} \tag{10-22}$$

Once the universe has crushed the sine waves into square waves at maximum compression, they will start to increase in amplitude and energy. This will cause a slight upward spike in light speed and the result is the big bang.

As the universe expands in all directions, in an Einsteinian space-time universe, a point is reached where there is no more expansion energy. The square waves have slowly converted into all sine waves. At this point we have reached full expansion. As we move a little further, the light speed has a slight downward spike. This destroys the protons, electrons, and everything else in the universe. The universe has now reverted to pure dots.

Once the universe has become a pure sine wave at half the energy of the big bang shortly after the big bang, the constant amplitude sine wave will slowly drop in amplitude. Eventually the amplitude will drop to near zero.

We see that the universe cycles from big bang to little bang as a standard electrical frequency modulated Doppler circuit. We now have two different methods of describing the universe. In both the GG/MM/Doppler and the F/M Doppler, the relative mass of the universe at big bang is twice that as of the present.

We see that a pure stationary mass looks like a single frequency. We also see that a moving mass becomes two frequencies and these two frequencies change from sine waves slowly to square waves. The net result is that a moving mass looks like a dual infinite series of frequencies.

SECTION 10-3: THE FM DOPPLER UNIVERSE

The FM Doppler universe is a different universe than our ordinary perceptions conceive. We observe a physical universe in which we can visualize length, mass, volume, and time. The GG/MM/Doppler solution gives us a mechanical perspective of the universe. The FM Doppler solution is purely electrical.

You cannot build a house or an airplane with the FM Doppler solution. It can help you to understand the variation of time clocks with velocity. It can help you to understand the cycle time of the universe, the big bang, the little bang, the nature of light, and electro-magnetic waves. It is a model of the universe.

The FM Doppler universe appears in two parts. The first part is that of pure free space where we limit our velocities to 0.01C or less. Pure free space tends to be very classical. Rulers remain constant. Clocks do not change. Finally mass does not change either. Pure free space is invariant space-time. It is also the solution for GG/MM/ Doppler at low velocities. Once we have strong gravitational fields, high-speed orbital motions, cyclotrons, etc., we no longer have pure free space. There we see the equations of Einstein take shape. The universe is a mixture of pure free space and orbital motion.

Let us look at the moving clock in pure free space in the FM Doppler universe. To the left of the motion we find a lower Doppler clock frequency. To the right we find a higher Doppler clock frequency. If we take the arithmetic mean frequency we get:

$$f_o" = f_o \tag{10-23}$$

The arithmetic mean clock frequency is the exact same frequency as that of a stationary clock. For the FM Doppler Universe in pure free space, it is a purely simple classical solution.

Arithmetic time = Invariant (10-24)

In an electrical universe in pure space at low velocities, a time clock does not change with velocity. For orbital motion, the geometric mean time clock is:

$$f_O' = f_O [1- (V/C)^2]^{1/2} \qquad (10\text{-}25)$$

We see in equation 10-25 that the FM Doppler geometric mean time clock drops in frequency as the clock cycles along the electromagnetic/ gravitational field. The classical FM Doppler solution gives Einsteinian equations as well. If we look at this clock in orbital motion, we can imagine an orbital circle with standing waves at f_O in all directions. The clock flows upon this circle. This produces higher frequency waves to the right of motion and lower frequency waves to the left of motion. The orbital clock flows within the field. You can say that the basic original field is stationary and that the clock is pushing the standing wave circle to the right. Yet, the clock still is within the circle.

From an electrical perspective, the clock is a perturbation of the electromagnetic field. However another way of looking at orbital motion is that the clock is repeating itself over and over again. As it moves faster and faster, an image of the clock appears upon itself. This slows the clock. This is only true for cyclical or orbital motion. It is not true for linear transient motion. Classical physics principles are based upon singular events and pure space-time. Once we repeat ourselves over and over again, we have orbital space-time or Einsteinian physics.

Both classical physics and Einsteinian physics are true. Yet, we think in terms of classical physics and orbital electrical principles are a little strange until we understand the differences.

If the clock is stationary, then it is a pure sine wave. As it moves it becomes a Frequency Modulated dual frequency entity. A moving electron splits into two separate frequencies that are very close together. The same is true of the clock. It is not a simple singular frequency entity. It is a duality. In general it is a spectrum of frequencies rather than only two.

In addition, the pure sine wave slowly turns into square waves. As the energy rises we get a greater percentage of square wave-shape out of the FM duality. The duality for the electron helps explains dual photons of almost equal energy in the spectra. Yet, since it is a spectrum, it will produce the entire spectrum between the two frequencies.

Let us return to the Michelson/Morley experiment from an electrical FM Doppler perspective. We are now looking at it from the electrical universe. Previously we looked at it from the mechanical universe. In that universe we found that the X dimension shrunk as the velocity increased. This was discussed in detail in the GG/MM Doppler chapters. The mechanical universe can produce the rings of Saturn and very interesting shapes.

The Electrical Universe cannot readily describe the rings of Saturn. Once we go into Fourier transforms and infinite series of wave-shapes, things get too complicated for our senses. The electrical universe and the mechanical universe are identical but many things are easier to describe mechanically. This is especially so because to build a house we think mechanically. The equations for a piece of lumber in electrical terms would be nearly impossible to understand.

Returning to the M/M experiment, when the equations were written, the time of travel in the back and forth motion in the X direction was compared to that in the Y direction. A null of the two beams indicated that the X dimension shrunk relative to the Y dimension. The basis of the null was that a sine wave of light hit orthogonal moving walls and returned to the measuring spot exactly in phase.

In this study, the assumption was that we are dealing with a consistent non-varying wavelength. The fallacy of this method is that the FM Doppler solution denies that the wavelength of the light is unchanging. For the null, the only thing important is that the phase angle of the two waves exactly matches when they hit the null point.

When we look at the experiment from the FM Doppler universe, the experiment is quite different. Let us look at an object moving toward the right. The clock frequencies are exaggerated in the figure below:

x   x   x   x[x   x   x   x x x x   xxxxx]xxxxxxxxxxxx   Fig. 1

In Figure 1, the clock is moving to the right. The Doppler frequency to the left is low. Within the moving clock, we go from low frequency at the left to an average frequency in the middle and a high frequency to the right. On the outside, the high frequency remains.

If we have a photon perturbation coming from the right, it will enter the M/M apparatus and will start to see decreasing frequency as it flows along the gravitational field. It will then rebound and start back to the right. The time spent on the right will take longer. However, it is even worse than the Einsteinium equations represent. Not only will the time back take longer but also more time will be spent at higher frequencies.

When Einstein took the simple calculation for the special relativity, which matches the M/M apparatus, he used a constant frequency basis. The equations are clearly different when we use FM Doppler. The same is true in the perpendicular direction. The photon perturbation starts upward at medium frequency but it ends up at the high end when it returns. This damages the simple equations Einstein used to derive his space-time.

The entire Einsteinian derivation of space-time was based on simple assumptions. In spite of this, when you deal with multi-frequencies, spectrums, etc., the root mean square method is prevalent in electrical theory.

The real universe is FM Doppler space-time. The equations are more complex. You have to deal with dual frequencies. You have to deal with sine waves that turn into square waves. You have to deal with infinite series. The glass within the M/M experiment tends to filter the higher frequencies. The square waves of light become sine waves when you filter it by glass.

When we look at the Universe in free space from the FM Doppler perspective, we find that:

$$\text{Arithmetic Distance} = \text{Invariant} \qquad (10\text{-}26)$$

In the electrical universe in free space and for velocities less than 0.01C, the arms of the M/M apparatus are invariant. The light in the free space FM Doppler universe moves over perpendicular arms in pure equal time. In the FM Doppler universe both distance and time are perfect. The same is true of mass. In the free space FM Doppler Universe:

$$\text{Arithmetic Mass} = \text{Invariant} \tag{10-27}$$

For the pure electrical universe, there is no change in mass with velocity. Einstein's equations for increased mass with velocity are quite meaningless in free space.

Finally for the FM Doppler Universe, the rest energy is that of a sine wave and the energy when moving at the speed of light is a square wave. Thus:

$$\text{Energy @ Light-speed} = 2 \text{ Rest Energy} \tag{10-28}$$

The FM Doppler universe is one possible model of the universe in which we live. It is a strange universe to us. In many respects it is quite classical. It can take the shape of a perfect sphere that expands and contracts.

It could also be a complex space-time sphere. The electrical universe can be flexible in shape. One solution would be a perfect sphere with us near the center. Another solution would be a complex wrapped space-time entity that behaves like a perfect sphere of radius R_U everywhere. Both solutions are possible.

Another possibility is that of an infinite universe in which every dot is defined up to the particular radius R_U. We then get overlapping space and time with particular areas oscillating from big bang to little bang while other areas are operating upon a different time sequence. This produces a very complex universe.

The FM Doppler universe can come from infinity as a sine wave of almost zero frequency. As it contracts, it still remains a sine-wave for a long time. Eventually it starts to become part sine-wave/part square-wave. Then it converts into a square-wave. The square-wave becomes a higher and higher frequency square-wave. Finally the amplitude of the square-wave increases at the big bang.

We will go from minimum radius at big bang to maximum radius at little bang due to the light-speed hysteresis loop. The light speed hysteresis loop is not part of the FM Doppler model. All that can be seen from the model is that the universe will compress into square waves and these will compress into higher and higher frequencies. At the other end the sine waves will stretch out toward infinity. The FM Doppler model does not permit the minimum radius of the universe calculation.

The beauty of the FM Doppler solution is that it is easy to see what is happening in the universe without struggling hard to understand. The electrical FM Doppler solution is very easy to understand.

SECTION 10-4: PHASE TIME

When we look at the FM Doppler Universe with non-varying time clocks, distances, and masses, the concept of phase time becomes very important. Phase time is the time to achieve the same phase angle on a sine wave. If we start at 30 degrees on a sine wave, we want to return to 30 degrees for our measurement.

For a constant frequency mode, phase time equals time. Thus:

Phase Time = Time (10-29)

In equation 10-29, if we have a constant frequency source such as a 60-hertz power generator, the time from 30 degrees to 30 degrees is constant. Thus:

Cycle time = Constant (10-30)

FM radio and later radar produced systems in which the frequency itself varied. Phase-time was born and phase-locked loops came into existence. Now we were locking on the phase angle of the signal. We no longer had constant cycle time but variable cycle time from cycle to cycle. The phase locked loop enabled meaningful data from a world of variable frequency time clocks.

Let us return once more to the Michelson/Morley experiment from the perspective of phase time. First let us look at a stationary frequency source of frequency f_o as read by a moving instrument with velocity V.

[[[f_o]]]   (<<<<< [V]<<<<)   Fig. 1

In Figure 1 we have a stationary frequency source [[[f_o]]] and a moving measuring instrument of velocity (V).   (<<<<<<<[V]<<<<<<)

The instrument reads:

$f_o' = f_o [1+ (V/C)]$ (10-31)

We see that the moving instrument reads a higher frequency than a corresponding stationary instrument. The instrument rides the electrical patterns of the stationary source and thus the phase time keeps getting less and less as the velocity increases. Let us now look at a stationary measuring instrument and a stationary frequency source of $(f_o)[1+(V/C)]$.

[[[$f_o\{1+(V/C)\}$]]]   [[[{V=0}]]]   Fig.2

In Figure 2 we have a higher frequency source and a stationary instrument. The instrument reads:

$f_o'' = f_o [1+(V/C)]$ (10-32)

We see that:

$f_o' = f_o''$ (10-33)

In Equation 10-33 we see that f_o' equals f_o''. There is no difference in measurement if a frequency source is coming toward a measuring instrument or a measuring instrument is moving toward the frequency source. This is for the pure electrical universe where frequency is constant and length and mass are constant as well. In the mechanical universe we will pick up a RMS Doppler for the moving clock, or a slowing of the test instruments time base for the moving instrument.

This does not occur in the pure electrical universe. We see that in the electrical universe, increased velocity is identical with increased frequency or decreased phase time when objects approach. When objects move away from each other, increased velocity results in decreased frequency or increased phase time.

Let us look at the M/M apparatus again:

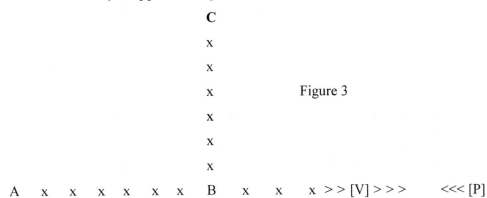

Figure 3

In Figure 3 photon [P] is moving to the left. The M/M instrument is moving to the right with velocity [V]. The photons hit the mirror at B, and split. Some travel to A and then back to B, others travel to C and then back to B.

In the usual derivation of the Einsteinian equations, the photons were considered constant frequency devices or a singular time line. Even though true photons are square-waves in pure space, the minute you hit the glass, you get a severe filtering of the higher frequency harmonics. The glass at B destroys the square waves and the result is sine waves.

The instrument was set to null for a stationary light and a stationary instrument. The dimensions were corrected to give proper results for a wide band of frequencies. If we went down to 60 hertz, it would be very difficult to produce any kind of null.

The wavelength would just be larger than the instrument. However we are using light so incoming frequencies come in multiple wavelengths and we can null the instrument for any frequency in that range.

If we had a frequency source inside the M/M instrument, we would find that the frequency coming out of it to the right would be higher and the frequency coming out of it to the left would be lower. We would find that the average frequency would be constant and that the root mean square frequency would be the rms Doppler or Einsteinian value.

We can then replace the instrument with a variable frequency generator, which looks like two frequencies. Yet it is really a variable frequency spectrum rather than only two.

$f_A[f_A\text{----------}f_O\text{---}f_O'\text{--------}f_B]f_B$ Fig.4

In figure 4:

$$f_A = f_O[1-(V/C)] \quad (10\text{-}34)$$

$$f_B = f_O[1+(V/C)] \quad (10\text{-}35)$$

$$f_O' = f_O[1-(V/C)^2]^{1/2} \quad (10\text{-}36)$$

Frequency f_O' is shifted to the right of (f_O) which is dead center. This is the Doppler root mean square frequency. It is also the Doppler center of mass shift.

The frequency (f_A) occurs both within the instrument and just outside it. The frequency (f_B) occurs within the instrument at the right side and also just outside it.

If the instrument is orbiting the Earth, standing wave patterns will be produced all around the Earth and both (f_O) and (f_O') will appear on the opposite side of the orbit.

Within the instrument itself we move from f_A to f_O and from f_O to f_B in a linear FM manner. We have a spectrum of frequencies. In addition sine waves turn slightly into square waves as the M/M instrument moves. We have a spectrum of visible light frequencies and a higher infinite spectrum of ultra high frequencies.

When the photons enter the instrument at point B, they enter a strange world of variable frequency. This is identical to the world of variable velocity moving along an electromagnetic field. The net result is that the phase time from B to A is identical with the phase time from A to B. Thus:

$$T_{P(A-B)} = T_{P(B-A)} \qquad (10\text{-}37)$$

In Equation 10-37 we see that the time to reach a particular phase angle on a sine wave such as 37 degrees is independent of velocity for a FM system. The same is true of the time to go from B to C. Thus:

$$T_{P(B-C)} = T_{P(C-B)} \qquad (10\text{-}38)$$

In equation 10-38 we see that the time to go from B to C is identical with the time to go from C to B. Equation 10-38 was never a problem for Einstein. It is only equation 10-37 that appears strange to our normal way of thinking with respect to constant frequency time clocks. The FM phase locked loop clock appears strange.

The most important equation is:

$$T_{P(A-B)} = T_{P(B-A)} = T_{P(B-C)} = T_{P(C-B)} \qquad (10\text{-}39)$$

In equation 10-39, we see that the perpendicular times and the horizontal times are identical. The instrument null is independent of velocity. The M/M instrument is a self-nulling device for a FM system.

The M/M instrument produced an optical phase locked loop. Michelson/Morley built a mechanical phase locked loop. Once it was set to null, it would always null. If the signals coming in are FM waves, it will act like a phase locked loop and null on them. We can get one set of equations from the instrument where time and distance appear invariant.

Likewise we can get the Einsteinian equations for the M/M instrument. Finally we can adapt the equations for the GG/MM/Doppler solution. It seems that we can look at the universe from several different viewpoints. The Einsteinian solution is one solution that works well for orbital motion. We then have two Doppler solutions that yield similar results to Einstein in one case and classical physics in another case. All this shows is that there are several different describing functions possible for the universe.

CHAPTER 11: THE LAWS OF GRAVITY

SECTION 11-0: INTRODUCTION

In this chapter, the gravitational field will be analyzed and laws will be introduced which explain the workings of the field. In addition, the equations of gravity will be explained from several viewpoints.

SECTION 11-1: THREE DIMENSIONAL FORCES IN SPACE TIME

Let us look at an ordinary object sitting in free space at zero velocity. The object exists all the way to the radius of the universe. If we look at the object as spectrums of mass-frequencies, we observe that waves radiate outward from the object reaching to the radius of the universe. We can also say that waves radiate inward toward the object from the radius of the universe. This will produce spherical standing wave forces upon the object in all directions. Since the object is perfectly stationary, the forces on the object are equal in all directions.

Let us now look inside the object. External gravitational forces are applied all around the outside of the object compressing the object. Forces all around the inside of the object pressing outward must counterbalance this. Let us look at the X-axis of the object.

$$F_{Ax} >[< F_{A'x} \text{ xxxxxxxxoxxxxxxxx } F_{R'x} >] < F_{Rx} \qquad \text{Figure 11-1}$$

In Figure 11-1 we have an object of spherical or cubic shape in which external forces are applied to the left as F_{Ax} and to the right as F_{Rx} along the X-axis. Corresponding forces are applied along the other two axes as well. These external forces are counterbalanced by internal forces $F_{A'x}$ and $F_{R'x}$. Thus:

$$F_{Ax} = F_{A'x} \qquad (11\text{-}1)$$
$$F_{R'x} = F_{Rx} \qquad (11\text{-}2)$$
$$F_{Ax} = F_{Rx} \qquad (11\text{-}3)$$
$$F_{A'x} = F_{R'x} \qquad (11\text{-}4)$$

In equation 11-1 we see that the external force on the left side equals the internal force on the left side. The same thing is true for the right side as shown by equation 11-2. In equation 11-3 we see that both external forces are equal to each other. Finally in equation 11-4 we see that both internal forces are equal to each other.

We can then write the equations for the Y-axis and Z-axis:

$$F_{AY} = F_{RY} \qquad (11\text{-}5)$$
$$F_{AZ} = F_{RZ} \qquad (11\text{-}6)$$

In equations 11-5 and 11-6 only one Y-axis equation and one Z-axis equation are shown. However, each axis has four equations just like the X-axis. The above equations are for an object at perfect rest. All the forces equal out. We have a stationary object at perfect rest. Let us now move the object in the X direction by applying a force F_{Qx}.

The initial application of the force F_{Qx} causes an unbalanced situation in the X direction forces. The object compresses heavily on the left side and less and less as we move to the right. We can look at the shape of the compression as a triangle. The compression density will be approximately twice as much on the left as the middle and zero at the right. This is the initial reaction to the force.

The compression will cause the center of gravity of the object to shift to the left. This is shown in Figure 11-2.

$$F_{Qx} + F_{Ax} > [\text{xxxxxxoxxxxxxxxxxxxxx}] < F_{Rx} \qquad \text{Figure 11-2}$$

In Figure 11-2 the C.G. at (**o**) has shifted to the left due to the addition of the applied force F_{Qx}. The force at the right does not change instantly. There is a transient condition where an unbalanced force occurs. This causes the object to start to move to the right. As the object starts to move to the right, the internal molecules of the object will start to compress to the right. We then get compression on the left and compression on the right and the object shrinks. In addition the center of gravity of the object starts to return to normal.

As the object moves the external Doppler forces on the right are in compression. This increases the initial external force F_{Rx} on the right. It also reduces the initial external force F_{Ax} on the left. The center of gravity of the object remains on the left side until we reduce the driving force. As the driving force F_{Qx} is reduced to zero we end up with no applied force on the object. However we are left with unbalanced Doppler forces.

The space-time C.G. shifts slightly to the right of center as shown in Figure 11-3.

$$F_A - F_{DA} > [\text{xxxxxxxxxxxxxxoxxxxxxxxxxx}] < F_R + F_{DR} \qquad \text{Figure 11-3}$$

In Figure 11-3 we see that the center of gravity at (**o**) of the moving object has shifted to the right. The initial equal forces on the left and right are no longer equal. A Doppler shift force appears on the left. This subtracts from the normal gravitational/inertial field forces acting on the object. On the right side we have a Doppler shift in compression which adds to the force on the right. The object is moving to the right but the forces are unbalanced.

The Doppler forces depend upon V/C. The differential force to slow the object is small but important. If we look at the force equation:

$$F = (d/dt)\, MV = M\, (dV/dt) + V\, (dM/dt) \qquad (11\text{-}7)$$

In equation 11-7 we see that if a moving object has a force acting upon it which opposes its motion, it will lose velocity or mass or both. Since the Doppler mass produces an unbalanced force, we have this rule of physics:

RULE OF MOTION: An object in motion with velocity V moving in a straight line will slowly radiate mass/energy over a long period of time until it finally breaks apart and stops.

Therefore, the law of motion in classical physics is wrong. An object will not continue in a straight line forever with constant velocity. The Doppler pressure will force every object in the universe to eventually stop. All the galaxies, all the planets, all the photons as well will eventually radiate their energy and stop. We see that another cause of the red shift can be conceived as the result of this rule of motion. Energy is lost and radiated as an object moves over long periods of time. The AC oscillation of the photon slows down over long periods of time. In the same light, the energy and mass of matter including the electron, proton, and neutron slow down over time. Therefore the electron and proton also have a red shift.

PROTON RED SHIFT: The proton, electron, and neutron have internal AC oscillation reductions over time. They experience the same red shift phenomenon, as does the photon. Their oscillation will eventually stop and they will break apart at the same time in the future, as will the photons.

If we look only at the X-axis we do not see the entire picture. Let us look at the Y dimension as the object moves in the X direction. When the object is stationary, the force diagram is the same as for the X direction. Thus:

$$F_{AY} > [<F_{A'Y}\ yyyyyyyyyyyyyyyoyyyyyyyyyyyyyyyF_{R'Y}>] < F_{RY} \qquad \text{Figure 11-4}$$

In Figure 11-4 we are looking at the Y-axis. The object is stationary in space. We notice that all the forces are equal. Now let us move the object in the X direction. The Y forces basically remain constant. The Y dimension of the object remains the same. The Y dimension center of gravity remains the same but the moment of Inertia increases in the Y dimension as the object speed up. This is because of the Doppler length increase in the X direction. If the object were rotating around a sun, this would cause the object to move into a higher orbit. The following chart illustrates the effect of X velocity motion.

Dimension	Center of Gravity	Moment of Inertia
Y dimension	unchanged	higher
Z dimension	unchanged	higher

We see that when an object moves in the X dimension, we get space-time changes in the moment of inertia of the object. The object gets inertially longer in the direction of motion and this causes the Doppler mass to move forward. The result is more inertia in the Y and Z direction. This will cause the object to move toward a higher orbit when revolving around a massive body.

SECTION 11-2: FORCES ACTING ON BODIES

Let us look at the various forces in action upon bodies in motion. If we apply a force to a mass we produce acceleration. The acceleration in turn produces a counter force due to the inertia of the object.

Although in classical physics, the counter force equals the applied force, if this was true, then the object would cease accelerating. The only way for the object to continue to accelerate is if there is a time delay between the applied force and the counter-force. Thus:

$$\text{Counter-force} = \text{Applied Force} \ @ \ \text{Time delay} \qquad (11\text{-}8)$$

In equation 11-8 we have an object in pure free space. For this condition, the only thing preventing the motion of the object is the self-inertia of the object. In pure free space, it does not take as much energy to bring an object up almost to the speed of light as upon the planet Earth.

Within a galaxy we can approximate pure free space in some places. We can accelerate objects quite rapidly when the gravitational field is quite low. Upon our Earth and in our surroundings, the Earth's gravitational field, the sun's gravitational field, and finally the galaxy's gravitational field effect us. When we move an object, the object's electrical fields cut all the gravitational fields and reactions occur. We do not get a single counter-force but a spectrum of counter-forces due to gravitational field eddy currents. We generate heat.

We never get a simple answer. We always get complex terms. However, depending upon the speeds and conditions some terms are primary and others are small or negligible. Whenever we move anything we have moving electrons, protons, neutrons, photons, Doppler mass effects, gravitational field effects, and eddy currents.

Einstein's variation of mass/energy with velocity provides us with answers for orbital motion. It is a particular case of pure orbital motion. Once we add linear motion to pure orbital motion we get complexities. One experiment can produce one set of answers and everyone will think Einstein is correct. Another experiment can produce another set of answers and everyone will think Einstein is wrong.

The truth is that Einstein's equations provide us with some good answers for some problems. In order to solve all the problems, one has to understand all the forces involved. Of course that is not too easy to accomplish so only the salient forces will be discussed. For some problems an insignificant force can become the most important force for that problem. Let us look at the forces within the Bohr atom again from a more detailed viewpoint.

There are three main force vectors pulling the electron toward the proton. There are four main force vectors repelling the electron from the proton. Some of the vectors are reactions to actions. The centrifugal forces tend to be reactions to the centripetal forces. Three of the vectors are centripetal forces and four of the vectors are centrifugal forces.

There are also force vectors associated with the spin of the electron and the spin of the proton as well. These will interact with the orbital vectors. There will be DC components and spectrums of AC components within all these vectors. The problem gets quite complex. However, most things will follow simple mathematical equations. Now, let us look at the forces from an elementary viewpoint.

The sum total of the centripetal vectors equals the sum totals of the centrifugal vectors although the individual parts may not be exactly equal to each other. In addition all parts have there own phase angles and or time delays associated with them.

The first centripetal vector is the coulomb attraction vector. This causes the main centrifugal force vectors to appear. Thus:

$$KQQ/R^2 = [MV^2]/R @ \angle A_1 + [U_O [QV/R]^2]/2\pi @ \angle A_2 \qquad (11\text{-}9)$$

In equation 11-9 we have the main coulomb attraction force equal to the sum of the centrifugal force, due to the rotation of the electron mass, and the magnetic repulsive force due to the flow of the electrons charge in the circular path.

Each term has their individual vector angle. The first term on the left side is chosen to have an angle of zero degrees for reference. We can now add a gravitational centripetal force to the left side. The motion of the electron through space will cause an anti-gravitational force on the right side. This will be a reaction to the gravitational force.

These forces are small for the Bohr orbit but will be significant for planetary motion. Thus:

$$GM_PM_E/R^2 \ @\angle B_1 = [(\Delta M)(V^2)/R] \cdot [\cos \angle B_2 + j\sin \angle B_2] \qquad (11\text{-}10)$$

In equation 11-10 we find that the gravitational centripetal force is counterbalanced by two centrifugal forces at right angles to each other. The forces appear as a change in the mass of the electron. This is for the simple case where the various parts are assumed to match each other. In the more complex case we have to add equations 11-9 and 11-10 together.

The gravitational forces are very small for the Bohr orbits. However, they have a small effect on the photon absorption spectra. Let us now look at the Einsteinian mass increase with speed. In the final neutron orbit it is significant. In the first Bohr orbit, it only produces 13.6 EV of mass/energy.

The minute we go into high Einsteinian orbital velocities, we start to develop dot current loops. Plus dots will move one way and minus dots will move the other way. The entire space in the orbit will be filled with flowing plus and minus dots. The net result of plus dots moving in space one way and minus dots moving the other way is an attractive magnetic field. Einstein's mass increase is partially counterbalanced.

If the rings of Saturn are moving near the speed of light, the Einsteinian orbital mass increase will tend to send them into space. The same is true of a large sun spinning at high speed around another sun. Once we add Einsteinian mass/energy increases to a spinning sun we have trouble keeping the two objects together. However, a moving object produces an electrical photon current such that plus dots will flow one way and minus dots will flow the other way. These electro-photons fill the entire orbit. An equal and opposite force is developed to counterbalance or partially counter-balance the Einsteinian force. Thus:

$$U_O \, N_D \, I_{D+} \, I_{D-} \ @\angle C_1 = [(\Delta M)(V^2)/R] \cdot [\cos\angle C_2 + j\sin\angle C_2] \qquad (11\text{-}11)$$

In equation 11-11 we have set the Einsteinian mass increase equal to the orbital current flow counter-force. The mass/energy increases, but it does not have such a dramatic effect due to the counter-force produced by dot current flow.

The above equations list the forces and counter-forces separately. However, this is only a special case where this is true. In general we must add all the vectors on the left side together and all the vectors on the right side must match the sum total. This makes the problem very complex.

Once we add the spin of the electron and the spin and motion of the proton, we have a very complex problem. Modern computer analysis should be able to determine all the forces in action in the Bohr orbit. Quantum physics will then not be necessary, as the actual details will become known.

The Dot theory and Doppler space time has extended classical physics into everything. It enables the calculation of things, which could not be calculated long ago when only two forces were in operation in the atom. Quantum physicists felt that a mathematical approach was necessary to do the job. Yet, in truth, it is merely complex electrical engineering that is required.

If an object is spinning very rapidly near the speed of light, the simple gravitational attraction and the simple centrifugal force no longer applies. It is only the plus/minus dot orbital current flow, which keeps the object in stable operation. We see that some forces apply at slow speeds while others take over at high speeds. In between we have a mixture of several forces. In general most problems only require two or three forces for their solution.

The orbital forces help us to understand many things. If we take an electron in the cyclotron and move it rapidly, it will spin quite fast. The spinning electron will produce dot current flows, which add to the normal electron spin. The minus dots will spin in the same direction as the excess minus dots and the positive dots will spin in the opposite direction. This will cause a strong magnetic field in front and behind the electron. It will also cause the electron to shrink toward a line.

In addition to this, the whole cyclotron will be filled with plus dots, which move rearward in the direction of motion and minus dots which, move in the same direction as the electron. This is due to the split of the photon energy. These lines of moving charge produce attractive fields. This will cause the electron to shrink into a line. The entire thing will be a line of charge, some of which comes from the original charge and some of which comes from the dot current flow both frontward and rearwards caused by the photons.

The dot current flow applies to all orbital situations. The Earth revolves around the sun. There are plus electro-photons moving in orbit around the sun in one direction and minus electro-photons moving in orbit around the sun in the other direction.

These currents produce forces, which counteract the Einsteinian mass buildup due to the Earth's velocity. These magnetic fields are attractive to each other on the same side of the orbit and have no net attraction or repulsion across to the other side of the orbit. They are like a contracting band or skin effect. A counter-force is built up which adds to the gravitational force from sun to Earth.

A side benefit of this counter-force is that we get a magnetic field perpendicular to the plane of rotation of the Earth. The electrically neutral Earth produces a net magnetic field due to rotation. The following law of orbital magnetism applies:

LAW OF ORBITAL MAGNETISM: An orbiting neutral body produces a magnetic field in an axis perpendicular to the plane of rotation.

The Earth in revolving around the sun will produce a magnetic field perpendicular to the orbital plane of the sun. This will be an orbital magnetic vector. The Earth as it spins on its axis will also produce a spin magnetic vector. The following law of spin magnetism applies:

LAW OF SPIN MAGNETISM: A spinning neutral body produces a magnetic field in the axis of spin.

The Earth's magnetic field has two components. The main component is due to the spin of the Earth on its axis. The second component is due to the orbit of the Earth around the sun. All of this is due to plus dot currents flowing one way and minus dot currents flowing the opposite way.

The Earth's magnetic field is a vector sum of the two fields. There will be reactions to these fields as well. In addition, large iron deposits unevenly placed within the Earth will change the magnetic path.

All the planetary motions have to include magnetic forces due to spin motion and orbital motion. The moon has one face locked with us but it still spins on its axis. The spin is just synchronized with the orbital motion. The magnetic fields produced by the spin and orbit help to keep the moon and Earth synchronized.

The Dot theory helps us to understand how neutral bodies can spin and produce magnetic fields. If you take a lump of aluminum and spin it rapidly it will develop a magnetic field apart from the effect due to the Earth's field.

Let us return to the hydrogen atom in Chapter 4. The velocity of the Bohr orbit in outward expansion was:

$$V_B^* = 1.05367 \text{ E-28} \tag{11-12}$$

Equation 11-12 is a repeat of equation 2-50, in which the velocity of the Bohr radius was solved from equation 2-49, which is repeated below.

$$G M_H M_H / R^2 = [2U_O QC / 137.036] \times [4 \pi Q V_B^*] / R^2 \tag{11-13}$$

From Equation 2-64 we get:

$$G = 16 \pi e U_O / (137.036)^3 \tag{11-14}$$

Combining Equations 11-13 & 14 we get:

$$M_H M_H = QC Q V_B^* (137.036)^2 / 2e \tag{11-15}$$

$$M_H = 137.036 Q [C V_B^* / 2e]^{1/2} \tag{11-16}$$

$$M_H = 58.772 Q [C V_B^*]^{1/2} \tag{11-17}$$

We see that the mass of the hydrogen atom (M_H) is equal to a numerical constant multiplied by the charge Q. This product is multiplied by the root mean square of the Bohr Orbit expansion velocity and the speed of light C. The numerical constant is the fine constant reciprocal divided by the square root of 2e.

Equation 11-17 tells us that mass is charge moving with a velocity V. In the case of the gravitational field, the expansion of the Bohr orbit produces the current flow which interacts with far away atoms to produce a net attractive magnetic field.

Let us relate the mass of the proton to the general Doppler formula:

$$M_P = 2 \pi Q_{OP} C \tag{11-18}$$

In equation 11-18 we have the mass of the proton equal to the RMS Doppler charge Q_{OP} times the speed of light. The 2π keep the equation the same form as the equation for the electron in Chapter 9 section 9-7. Solving for Q_{OP} we get:

$$Q_{OP} = M_P / 2\pi C = 8.87968 \text{E-37} \tag{11-19}$$

We now have a Doppler charge for the proton approximately 1836 times as much as for the electron. The proton and electron have the same local charge but the Doppler charge at a far distance of the proton is much larger than that of the electron.

This is another way of interpreting the mass difference. Mass is a Doppler property of charge. The proton carries a greater plus/minus dot current flow than the electron. The differential currents are the same and the voltages are the same but the proton acts like many more charges than the electron. Let us now look at the energy of the proton:

$$h f = h C / \lambda = \text{Energy} = \text{coulomb meters}^3 / \text{seconds}^3 \qquad (11\text{-}20)$$

In equation 11-20, λ is the wavelength. Dividing by C, we get:

$$h / \lambda = 2\pi Q_{OP} C^2 \qquad (11\text{-}21)$$

In Equation 11-21 the term h/λ is a momentum and $2\pi Q_{OP} C^2$ is a momentum term. The $2\pi Q_{OP} C$ is mass and the C is the speed of light. If this was a photon, it would be traveling at the speed of light and have an energy equal to:

$$\text{Energy} = 2\pi Q_{OP} C^3 \qquad (11\text{-}22)$$

From Equation 11-22 we see that the energy of a proton equals its Doppler charge multiplied by the speed of light cubed and times the factor 2π.

SECTION 11-3: THE GRAVITATIONAL FORCE

Let us look again at the equations for gravity and understand what they mean. Rearranging Equation 11-14 we have:

$$G = [U_0 / (137.036)^2] \cdot (16\pi e)/ 137.036 = 6.6720\text{E-}11 \qquad (11\text{-}23)$$

In equation 11-23, the gravitational constant is identical in units as the electrical permitivity constant except for the factor 137.036 squared and the correction factor of ($16\pi e / 137.036$). U_0 is a stronger term than G by the basic factor of 137.036 squared.

The 137.036 squared are the amount of cycles the electron travels in the first Bohr orbit to repeat itself. The electron moves basically 137 full waves for each time it circles the proton. Then it shifts 137 full waves in the perpendicular direction to form a surface pattern of 137 times 137 waves. The number 137.036 instead of 137 was due to the fact that the waves do not reach the exact same point but are shifted 360/274 degrees per revolution.

The electrical equation for the gravitational force from Equation 11-13 is:

$$F = [2U_0 QC/ 137.036] \cdot [4\pi Q V_B^*] / R^2 \qquad (11\text{-}24)$$

In equation 11-24 we see that the force of gravity is due to a charge Q moving at the Bohr orbit speed of C/ 137.036. This produces the term QC/137.036. This is a charge times velocity term or charge momentum. The force of gravity is also due to a charge Q moving at the Bohr orbit perpendicular expansion velocity V_B^*. We see that a 4π term also exists which tends to indicate that we are dealing with a surface effect. The expansion velocity V_B^* tends to act over an entire surface rather than a force between two straight lines. Let us rewrite Equation 11-14 in terms of U_0 and put this expression into equation 11-24. Thus:

$$F = 2 G (137.036)^2] \cdot [137.036/16\pi e] \cdot [QC/137.036] \cdot (4\pi Q V_B^*) / R^2 \qquad (11\text{-}25)$$

In equation 11-25 we see that the 137.036 squared factor magnifies a gravitational attraction from a singular attraction between an orbital electron current and the effect of an expansion velocity upon this current.

What does this equation mean? The gravitational field of this Earth can be looked upon as the field of single hydrogen atoms times the equivalent number of hydrogen atoms crushed together within the Earth. Basically the heavier elements produce more dense structures of equivalent numbers of hydrogen atoms. Except for binding energies and the like, an atom with 6 protons and 6 neutrons will have almost the same gravitational effect as 12 hydrogen atoms.

If we look at the Earth we find that the atoms are expanding and that the spaces between atoms are also expanding. The term QV_B^* can be considered a tiny current caused by the charge Q moving in an expansion direction. Alternatively, the term gives the difference in the position of the electron shell at the present time compared to its position a split milli-second ago thereby producing a surface attraction by Coulomb's law. A final alternative explanation is the Doppler gravitational force. This will be explained in chapter 12.

If we look at this current in three dimensions and place it aside another similar current, we find a point source current. This produces a point magnetic field as well. It is a strange type of current and is quite different than Ampere's laws. It is a current, which flows across the barrier. If we look at each section of the current flow from the point source and compare it with a second point source we find that the total attractive forces and repulsive forces nullify each other in the normal viewpoint because we are not used to spherical outward current flows. However spherical current flows do have attractive fields. Alternatively the force is like a compressed wave which attracts other compressed waves as they simultaneously expand.

Since binding energies tend to be local. The fact that the electron is expanding at the velocity V_B^* indicates a force which is common to the expansion of the universe itself. It is a space time force. It is a Doppler force.

Let us look at the ratio of the gravitational force to the electrical force. Thus:

$$GM_H M_H / KQQ = V_B^*/C \ [2.304946] \tag{11-26}$$

In Equation 11-26, The gravitational constant G=6.6726E-11, the mass of the hydrogen atom M_A=1.67353E-27, Coulombs constant K=8.89756E9, the charge Q= 1.60218E-19, the speed of light C = 2.99792E8 and V_B^*=1.05355E-28. The constant 2.304946 is close to the natural log of ten. Thus:

$$\ln 10 = 2.302585 \tag{11-27}$$

We see that to within 0.1 percent, the ratio of the gravitational force to the electrical force is the ratio of the Bohr expansion velocity to the speed of light and times the natural log of ten. Thus:

$$\text{Gravitational Force/ Electrical Force} = V_B^*/C \ \ln 10 \tag{11-28}$$

In equation 11-28, the ratio of V_B^*/C is a Doppler velocity ratio. If we look at the Bohr orbit or the Earth itself as a wave with a spectrum of frequencies, the gravitational force is due to the back-pressure from the universe caused by the expanding atom or Earth, which produces a Doppler increase in the wave frequencies.

Likewise, if we look at the universe as a pure electromagnetic field, and the Earth as standing wave vibrating within this field, the expansion of the Earth produces pressure on the external field. Standing waves from the radius of the universe inward toward us are Doppler compressed which causes small Doppler frequencies to radiate away from us.

$$f_D = f_O [V_B^*/C] \tag{11-29}$$

In equation 11-29 we see that the spectrum Doppler frequency differences (f_D) will be very tiny, where (f_D) represents a frequency spectrum. In linear motion the frequency to the left of motion is less than (f_O) and the frequency to the right of motion is larger than (f_O). In that case we get two different frequencies. In the case of the gravitational field, the Doppler frequencies are the same in all directions and radiate outward gravitationally with a back-pressure inward. In addition, the Earth's motion produces a velocity component with its associated Doppler frequencies.

In the case of small linear velocities the addition of momentum produces a net increase in the forward frequency and a net decrease in the rearward frequency. In the case of gravitational motion, the tiny Bohr orbit velocity has the same counter-effect as the forward frequency increase for linear motion. Both linear motion and gravitational expansion motion involve pressure upon the surface, which is moving in the forward direction.

The gravitational force then is a back-pressure from the universe upon the expanding atom or Earth. When we look at two atoms or two physical bodies separated a distance apart, we can see that pressure from the universe acting upon both tends to focus at the center of gravity of the two objects

Gravity then is the pushing together of bodies by the universe. The electromagnetic field pushes two bodies together.

If we can understand the forces of gravity on one hydrogen atom within itself, then we can see how the universe pushes upon the single atom. The force field then becomes the force per atom times the number of atoms. It depends upon the mass. In basic physics, the acceleration due to gravity on the surface of the Earth is:

$$A_G = G \, M_e / (R_E)^2 \tag{11-30}$$

In Equation 11-30, the mass of the Earth depends upon the number of equivalent hydrogen atoms times the mass of one hydrogen atom. If we understand the force on a single hydrogen atom, then gravity is understood. From wave theory, we basically understand that back-pressure upon the Earth will produce gravity. From the Dot theory we see that there is an outward flow of spherical dot current from the Earth toward the universe. It is like an expanding sphere of light.

Waves and particles are understandings of the physicists. They can produce wave and particle expressions for the gravitational field. Einstein produced curved space-time to express gravity. Particles and waves and curved space-time are all complex understandings of gravity. One can say that gravity is the result of waves. One can also say that the dots are the gravity sub-particles sought by physicists.

If we go back in time, we can say that the velocity of the Bohr Orbit or the Earth has an acceleration component. We can say that the velocity of the Bohr Orbit increases with time. This is especially true since in the common mode solution, the ruler is expanding with time and time is expanding with time. Both time and distance follow an exponential function and we can say that the universe has common mode acceleration in it. Thus:

$$A_{(Bohr)} = V_B^* / T_U \qquad (11\text{-}31)$$

In equation 11-31 we can define acceleration for the Bohr orbit by assuming that the Bohr velocity was initially zero and that the acceleration was constant up to this time. We could also say that the acceleration long ago was greater and that today it is less. Thus some of the force of the Earth upon our feet is due to the acceleration of the physical body of the Earth itself.

Since we can assume a general acceleration for the Bohr orbit, we reach the position of the man in Einstein's elevator. However, the acceleration of an elevator in outer-space devoid of any gravitational field is not the same as the acceleration of the Earth upon a person. The principles of transient space-time would apply to the elevator problem while steady state orbital space-time would apply to a similar problem within a moving gravitational field.

Let us now return to Equation 11-24. We see that we have a Bohr orbit electron, which is oscillating at 137 full waves in one direction and 137 full waves in the perpendicular direction per cycle. We see that the electron is moving outward spherically at the rate of V_B^*. This is an extremely small rate. We see that during the next cycle we repeat the same pattern of full waves. There will be a slight phase angle between patterns but for the most part, each set of waves is phase locked with the prior set of waves.

AC electrical currents flowing in the same direction are attractive if they are phase locked and basically in phase with each other. If the phase shift is ninety degrees, they have no attraction and if the phase shift is between 90 degrees and 180 degrees, they will repel.

Since the electron produces a basic standing wave pattern for each cycle and since the phase shift is near zero degrees, we have an attractive current pattern.

Although the electrons of different atoms within the Earth are not phase locked, all atoms have electrons, which are phase-locked with their own protons. All electrons spinning around a proton have a very tiny space-time force downward toward the proton. The same is true of the proton itself. It too has a very tiny space-time force toward its center due to the proton's AC oscillation as per chapter 5.

This spherical downward force is common to all atoms and is a spectrum of forces. The factor 4π in equation 11-24 shows that it is a surface force. Equation 11-24 tells us that an electrical force exists between the present cycle of the electron and it's past cycle. Gravity then is the link between the past and the present.

The ordinary electrical force of attraction between electrons are in the here and now. The gravitational force is still an electrical force but it is a space-time force. As the universe of today expands, it is pulled together by the universe of yesterday. Gravity is a general space-time property of the universe rather than a simple here and now electrical problem.

If we look at the present universe and compare it to the universe of a split second ago, we will find that it is larger. All over the universe forces will exist which tend to restore the universe, yet the universe will constantly expand. Eventually however, the protons and electrons will slow greatly and explode and the contraction cycle will begin.

If we take many atoms in a spherical configuration, the forces of gravity will be such that the individual vector forces will add together. The vector forces of today will focus toward the center of yesterday. In effect we can replace one million individual hydrogen atoms with a single equivalent hydrogen atom of one million times the mass and located at the center of gravity of the atoms.

The equivalent of two hydrogen atoms a distance R apart is a single hydrogen atom located halfway between them with twice the mass. This is only possible if the force of attraction between the currents of each cycle of the hydrogen atom encompass the entire universe.

If the origin of the electron only existed at the electron itself, then the local force between the cycles of the Bohr orbit spherical shells would not effect any other atom. The fact that two atoms have an equivalent half way between them indicates that the electron exists all over the universe. It is a focal point of a field.

As was pointed out in Chapter 9, the space-time image of each dot exists at the radius of the universe. Therefore each electron, proton, and neutron also exist at R_U. In effect the electron has a here and now image and an inverted image at the radius of the universe, R_U.

The earth is composed of many atoms. The total force field, which depends upon the density distribution, is basically at the center of the Earth. All the atoms tend to look like one atom at the center of the Earth. At the surface of the Earth is the large force field due to all the atoms of the Earth. We react with this force field. The field always points to the center of the Earth. It is an electrical force field.

In many respects the electrical gravitational field is a strange field. The field depends upon the attraction of the electron with the memory of itself and the property of the total field forming an equivalent field at the center of mass.

In wave theory, if we send two sets of waves outward from two different points, they will tend to reflect backward toward a common center point. If we look at each hydrogen atom we can say that there is a wave force field which moves outward and then reflects backward toward a common center point.

The gravitational force gives us the impression that the electromagnetic field itself appears as some sort of liquid ether. As we expand the atom, outward waves in the ether develop which reflect backwards toward us. It is always possible that the waves reach huge distances within a galaxy but eventually reflect backwards toward us without encompassing the entire universe.

In linear motion we push the electromagnetic field and get a reaction called inertia, which attempts to prevent the motion of the object. While the object is accelerating a force is built up. Later at constant velocity forces and counter-forces exist which are equal and opposite. Moving objects have built up electrical forces, which tend to cancel themselves.

With gravity, the forces and counter-forces are spherical in nature. We can assume a linear acceleration for the atom. Yet, we can also say that with gravity most things tend to remain in a stable situation with forces and counter-forces in balance. The Earth revolving around the sun exists with force and counter-force in balance.

Gravity can be thought of as a force of disturbance within a perfectly balanced electromagnetic field. Everything appears in perfect balance. The two hydrogen atoms at far distances appear as perfectly balanced fields. Locally, it takes some distance to nullify the plus and minus fields assuming they are exactly equal. A million miles away a hydrogen atom appears neutral. The plus and minus fields cancel perfectly. However, they both exist to the radius of the universe.

We have very strong electrical fields coming out of a single atom and a lump of matter. The two lead balls in the lab contain huge DC fields. These DC fields in turn can be looked upon as peaks and valleys in space and time or vibrations which are out of phase and with opposite rotations. Plus and plus repel and plus and minus attract. Therefore we have huge forces at work within the electrical fields.

Since the electrical forces are balanced, they tend to be invisible. They form the electrical ether of space and time. In Equation 11-24 we see that the electron is moving at $C/137.036$ within the balanced field. It is creating waves and disturbances within the field. In addition, the Bohr orbit is expanding at $V_B{}^*$ within the strong electromagnetic field. The motion of the electron within the strong balanced field is like a pebble thrown into a lake. The vibration of the electron alone is bad enough. Somehow the strong field can form standing waves around it and isolate it. However the electron is also expanding and taking up more space. It is putting more outward pressure upon the field.

Another atom stands nearby doing the same thing. They push on the strong field and the field pushes back on them. The center of gravity of the two atoms becomes the center of inward force of the strong balanced field. Gravity then is not an ordinary electrical attraction force. It is an internal force, which pushes on balanced fields, and these fields push back.

Since photons are balanced we can say that gravity is a photon field. It is the interaction of light waves with each other. An expanding light wave pushes on the universe and the universe pushes backwards. Likewise each hydrogen atom contacts with the electromagnetic field around it causing a pressure gradient.

We then have two understandings of gravity. From the electrical perspective, gravity is the electrical force between the universe of a split second ago and the universe of the present. The memory properties of space and time permit this tiny electrical force to exist. This is what equation 11-24 specifies.

We could also look at this equation and see that it represents the wave motion of the electron within the balanced electromagnetic field. We can interpret the same equation two different ways. Physicists could also write Doppler wave equations without reference to any electrical equations as well. Alternate understandings of gravity can be produced. This book provides one way of looking at gravity from the electrical perspective.

SECTION 11-4: GRAVITY IN SPACE TIME

The gravitational equation in Section 11-3 can best be understood as a space time equation. Let us repeat it here:

$$F = [2U_0 QC/137.036] \cdot [4\pi Q V_B{}^*] / R^2 \qquad (11\text{-}32)$$

Equation 11-32 is the force between two hydrogen atoms a distance R apart. The force can be looked upon as the force field of a hydrogen atom acting upon a test atom a distance R apart. The more atoms involved the greater the total force.

Equation 11-32 can be looked upon as a space-time force field. The force depends upon the motion of the electron at C/137, which is a current field. It also depends upon the outward velocity of the Bohr orbit. This determines the distance the Bohr orbit has moved each complete cycle, as the electron moves 360 degrees times 360 degrees around the proton producing a surface current.

The space-time force generated is the force of the present cycle, which is synchronized, with the memory of the previous cycle. Normally unsynchronized AC type fields have zero net attraction. Synchronized fields attract depending upon the cosine of the angle of synchronization. For the case of zero degrees, the attraction is one hundred percent.

The gravitational force field directed from the surface plane of the electron toward the proton's center is due to the tiny motion of the Bohr Orbit each cycle. If the Bohr velocity V_B^* increased, so would the distance between cycles. The force depends upon the distance traveled per one whole complete surface cycle.

The force can be looked upon as a pressure. At the surface of the electron, there is a downward pressure toward the proton. As we move toward outer-space, that downward pressure toward the center of the proton lessens by the square of the distance. In this case space looks similar to a substance or ether. Electrons and protons and the dots appear as imperfections in the uniform ether. Likewise if we view the dots and the electrons and protons as focal points of the electromagnetic field, it is the field itself which is expanding and producing the back-pressure upon the electrons. The field pushes them toward the protons as the electrons are moving away from the protons due to the universal expansion of the field. The force is a reaction to the expanding universe.

The force field of the atom itself is easy to visualize. This can be extended to the Earth itself. The force field of the Earth is merely the total field of all the atoms. Each multiple-proton atom is merely the equivalent of many hydrogen atoms. The field centers at the center of mass of the Earth. At the exact center of gravity of the Earth, the gravitational forces are balanced in all directions.

The individual gravitational pressure of a single atom adds to all the other surrounding atoms to form a field around the entire mass. When we look at the two lead balls in the laboratory, each lead ball has its gravitational force field centered within itself. It is always difficult to understand why the two lead balls attract each other. However, we must look at magnetic force fields and see the way two magnets tend to form circular field lines around each other, which minimize the length of the field lines. Two north/south magnets separated by a distance and aligned north to south and south to north tend to have a magnetic center halfway between their individual centers.

The gravitational force field of the two lead balls likewise tends to exist centered between each other. The two lead balls are pushed together by forces, which tend to shrink the far out field lines in a manner similar to the magnetic fields.

In addition, the past and present magnetic fields form spherical planes around the two objects, which push them together by the extension of equation 11-24 to the fields, themselves. Maxwell type equations could be written for every point in space.

The two lead balls do not in general have synchronized atoms. At the atomic level, the protons synchronize with each other to form local binding fields. When we take a huge number of atoms, we do not find that everything is synchronized. The synchronization of one group of atoms to form atomic bonds and then chemical bonds tends to negate other atoms a distance away. Except for magnetic substances, large groups of atoms tend to be magnetically neutral.

The gravitational field in the here and now between two lead balls is zero. There is no net attraction between the two lead balls due to the currents of this split second of one lead ball and that of the currents of the second lead ball. Every current flow and every electrostatic attraction and repulsion is met with an equal and opposite force.

The gravitational force is not an ordinary electrical circuit force. The equations of gravity look ordinary but they are not. Normal electrical theory relates a current in one wire and a current in a second wire. It also relates a current in a wire with the changing magnetic field of that wire. This is all here and now forces. The gravitational force relates the electrical attraction between the present and the past. It relates the current in the wire at the present with the previous current in the wire caused by the identical electron. It relates the current of today with the memory of the current of yesterday. Of course yesterday was only a split milli-second ago.

Gravity then is the force of the electron, proton, and neutron with their own image in the past. In general, the past, the present, and the future coexist for a split second of time. The universe would not hold together if this were not true. The memory properties of the universe are the cause of gravity. They are also the cause of the double slit phenomenon.

Equation 11-25 shows us in concrete terms that the only explanation for gravity is the force of attraction between the present and the past. Ordinary current flows will form equal and opposite patterns which nullify each other. The only electrical force, which will not nullify itself, is the perfectly synchronized force between one cycle and the next cycle of the complete Bohr orbit in all directions.

Now that we understand the space-time gravitational force, we can look at the inertial forces of a moving object. Let us look at a proton standing still in space. It contains a net electric field, which reaches outward toward the radius of the universe. This field is positive due to the excess positive dots over the negative dots. Contained within the proton are plus and minus dot current flows, which hold it together. The net forces of attraction are stronger than the net forces of repulsion and the proton will oscillate internally to balance out the forces. The proton will not readily fly apart unless caused to oscillate in an uncontrolled manner.

If we move to the very center of the proton, it will be similar to moving to the very center of the Earth. We will find that a test charge placed exactly in the center of the proton will have zero net force acting upon it. As we move the positive or negative test charge away from the center of the proton, the force will increase. The maximum force will occur at the surface of the proton. Thereupon, the force will decrease as we move away from the proton.

The electrostatic force within the proton can be written as:

$$F = [KQ^2 / (R_P)^2] \cdot (R / R_P) \tag{11-33}$$

In equation 11-33 the force acting upon a test charge Q of zero size within the proton is zero at the center and increases as we move to the radius of the proton R_P. The force outside the proton for the test charge is the standard coulomb force:

$$F = KQQ / R^2 \tag{11-34}$$

The second force within the proton is that due to the dot current flows. This is the magnetic force field of the proton. It is an AC spherical field, which forms standing waves in the vicinity of the proton. If we place another proton nearby the two AC fields will synchronize and a binding energy results. In general the magnetic field is approximately twice as strong as the electrostatic repulsive field. Two protons will bind magnetically with a force of twice the electric force. This produces a net force equal and opposite to the electric repulsive force. When we add many protons, they no longer can bind in pure phase. We then produce phase shifts or various angles and diamond has perfect 60 degrees phase shifts as previous discussed in Chapter 5.

We see that we have two forces in action when we push a proton. We have the electrical force and the stronger magnetic binding current force. When we push the proton we displace the center of charge of the proton with the memory of itself. We get a repulsive force between the proton and the memory of itself. Thus:

$$F = - [K_G Q^2 / (R_P)^2] \cdot (R/ R_P) \tag{11-35}$$

In equation 11-35, K_G is the equivalent gravitational electrical constant which is very small compared to Coulomb's constant K. We see that as we displace the proton from the image of itself, we get a small repulsive force as shown by the negative sign. The binding force currents within the proton produce a positive force of twice the amount as the electrical repulsive forces. Thus:

$$F = [2K_G Q^2 / (R_P)^2] \cdot (R/R_P) \tag{11-36}$$

The magnetic force is twice the electrostatic force. This produces a net force of magnetic attraction between the present and the past as we move the proton. Thus:

$$F_{(Net)} = [K_G Q^2 / (R_P)^2] \cdot (R/R_P) \tag{11-37}$$

The net force opposes the applied force. The inertial force is very similar to the gravitational force. In the gravitational force, a spherical force field existed which produced a force field in all directions. We look at this magnetically but we could also say that a repulsive electrostatic force exists as well. The magnetic attraction gravitational field is twice the strength of the electrostatic repulsive magnetic field.

We can look at the gravitational repulsive field by having the charge Q within one sphere and seeing the charge Q on the sphere of yesterday. This will cause a net repulsion between the sphere of yesterday and the sphere of today.

Alternately we can look at the charge in the Bohr Orbit shell and see repulsion between the shell of today and the shell of yesterday. The same would be true at the radius of the universe. There will be electrostatic repulsion between the electrostatic universe of today and the electrostatic universe of yesterday. The net force of gravity will be the difference between the strong electromagnetic field and the strong electrostatic field.

We can look at this as the driving force of the expansion of the universe as the charge Q slowly decreases in time. The magnetic effect becomes the counter-force, which attempts to prevent the expansion of the universe.

Therefore we could say that the electrostatic force and the magnetic force are really equal and opposite forces but have different effects. The electrostatic force expands or drives the universe but the magnetic force tends to bind the universe to prevent expansion.

In the Dot theory we had a capacitive/inductive oscillator. The dot charge becomes the dot current. At maximum radius the dot is basically discharged and the current is stored in the inductance of space.

The gravitational pump is the creation and discharge of the dot charges. Both the electron and proton expand as they discharge. Equation 11-32 is written in terms of the Bohr orbit. The orbit expands and the magnetic field attraction between past and present becomes self-evident. Within the proton and electron similar things happen. They too expand and have equal and opposite forces.

When we remove the applied force from the proton, the Einsteinian solution would restore the proton to normal. The past and present would return to itself. The electrostatic force between the past and present would be zero as would be the magnetic force. It would be independent of velocity. However, that is not true. A moving object depends upon a total velocity relative to the speed of light. When the force is removed, a permanent displacement between the past image and the present image remains. We then have a permanent repulsive force between the past and present of the proton. We also have a permanent attractive force between the past and present of the proton due to the binding current of yesterday and the binding current flow of today. For stability:

Force (electrostatic) = Force (magnetic) (11-38)

An object that is moving has equal and opposite force operating within it. The energy required to move the object is converted into the energy required to separate the object's past from its present. Once the object has gained velocity, there is a permanent electrostatic force tending to separate the object and a permanent magnetic force tending to bind the object.

We could say that when we move an object we modify the shape and cycle time of the binding currents. We could also say that we produce an additional circulating dot current flow in opposition to the binding current, which also opposes the motion.

The net result is that motion tends to produce directional magnetic fields within the atoms. If all fields took the same direction, a moving object would have a small magnetic field. Yet, the magnetic fields of some atoms may move in one direction and other atoms in the opposite direction. This will produce complex circulating currents within large objects.

The magnetic field of the Earth can be caused by simple vector currents due to rotation of the Earth and motion around the sun. However, we could get more complex circulating current flows. We could then get a net magnetic field, which depends upon the motions of the Earth and the distribution of the magnetic materials.

We see that if we take an electron and bring it up to light speed, we produce a situation in the lab where the electron is no longer contained within the radius of the electron. We then have a line of charge and a complex pattern of binding energy.

Einstein felt that each object was the center of its own universe. In many respects this appears to be true. When we look at the gravitational field, it comes from the object itself and the reaction of the object to the universe. In the Dot theory, the object or the dot is merely a focal point. The dot, and or the object have a field, which extends all over the universe.

When we push the object spherical electrostatic forces tend to produce repulsion between the past and the present while spherical magnetic forces tend to cause attraction between the past and the present. This is all happening locally at the object itself although the effects reach all over the universe.

If we quantize gravity, the effect of a single atom eventually dies out. It takes a huge sun to have much effect a billion miles away. Even a galaxy has little effect upon another galaxy, which is separated by huge distances.

We see that gravity is an electromagnetic effect. It is a simple force between the past and the present. We also see that when we push an object we build up electromagnetic forces, which oppose the motion. We also see that a moving object has permanent separation between the past and the present and permanent electrostatic and counter electromagnetic forces built up which contains the energy of motion.

The mystery of gravity has been solved. We now understand what happens when we push an object as well. In both cases we are dealing with the interaction between the electromagnetic field of today and that of yesterday where today and yesterday are only a split milli-second apart. In the next chapter we will look at gravity from a space-time viewpoint.

CHAPTER 12: UNIVERSAL GRAVITATION

SECTION 12-0: INTRODUCTION

In this chapter, the laws of universal gravitation will be described. In addition, the Doppler gravitational force will be analyzed.

SECTION 12-1: THE LAWS OF UNIVERSAL GRAVITATION

In this section, let us look at the reasons why the two lead balls in the laboratory attract each other. Let us look at a positive charge +Q moving to the left near the top and to the right near the bottom. Using the right hand, with the thumb representing the current flow, the magnetic field will be downward at the top and bottom and upward in the middle as shown by Figure 12- 1:

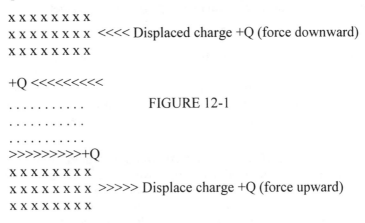

FIGURE 12-1

In Figure 12-1, the X's represent the downward magnetic field due to the motion of the charge +Q and the dots represent the upward magnetic field of the same charge +Q.

Originally, the charge +Q moves in the null center of the magnetic field in each direction as shown by the symbol (+Q<<<<<<<) or the symbol (>>>>>>>>>+Q). An electron moving around a proton will produce an inner magnetic field between the electron and the proton and an outer magnetic field moving outward toward the radius of the universe. The electron will flow along a neutral path between these segments of the field.

As the electron rotates 360 x 360 degrees, the magnetic field at any point will change. Inner and outer fields will turn into each other and the static DC field will be an AC field. The result in these fields will be standing wave patterns.

Two hydrogen atoms at a far distance will find themselves with attractive and then repulsive fields due to the two electrical charges and the magnetic effects as well. This will cause oscillations of the atoms. They will attract and repel and oscillate. If we take large groups of atoms, they will form phase angles within molecules and form complex patterns with oscillating attractive and repulsive electrostatic and magnetic fields. Everywhere standing-wave patterns will be produced, yet, the individual plus and minus charges and individual magnetic fields tend to exist all over the universe.

The Earth produces oscillating patterns of electromagnetic fields, which reach to the moon. These complex patterns could possibly have a net attraction or a net repulsion or be purely neutral. The AC field patterns do have full electromagnetic field strength. Two hydrogen atoms have large plus and minus electrostatic forces and large magnetic forces as well. The force between the Earth and the moon is:

$$\text{Force} = (+/-)\text{Net Electrostatic force} (+/-) \text{Net Magnetic force} + F \text{ gravity} \quad (12\text{-}1)$$

Equation 12-1 tells us that the force between the Earth and the moon is due to a net electrostatic force attraction or repulsion plus or minus a net magnetic force attraction or repulsion plus a gravitational force.

The gravitational force itself relates to the dynamic difference between the net electrostatic force and the net magnetic force. These are all complex AC type forces and there is no easy way of calculating the effect of a spectrum of billions upon billions of atoms upon the Earth interacting with huge amounts of the same upon the moon. Therefore we must look only at a single atom and see how the gravitational field works. Equation 11-24 gives us a very simple equation for the gravitational force. Rearranging the terms we get:

$$F_G = (8\pi\, U_O\, QC/\, 137.036) \cdot QV_B^* / R^2 \quad (12\text{-}2)$$

In equation 12-2 we see that the force between two hydrogen atoms depends upon the Bohr Orbit velocity ($C/137.036$) and the expansion of the Bohr Orbit due to the small velocity V_B^*. The gravitational force between the Earth and moon depends upon the total number of atoms within the Earth and moon, the Bohr velocity, the expansion of the Bohr orbit, and the distance from the Earth to Moon squared.

The complexity of the DC and AC magnetic fields between the Earth and moon tend to provide an oscillating electromagnetic field which acts like a washing machine or agitator. The Earth and moon exist within a very complex electromagnetic field, yet this only tends to agitate them. They will neither attract nor repel each other due to the electromagnetic washing machine agitation. However, once we can produce a force vector from Earth to moon and from moon to Earth, the washing machine will do the rest.

The powerful electromagnetic fields, which appear to be balanced between the Earth and moon, will provide the power to push them toward each other. The fields are very complex but within the fields are the simple space-time gravitational fields.

Let us return to Figure 12-1. A charge Q is moving to the left in the upper part of the figure and to the right in the lower part of the figure. There is no net magnetic field force upon the charge in either the upper or lower path. Let us now expand the universe, the magnetic field represents the past and the position of the charge represents the present. The charge has been displaced upward at the top of Figure 12-1. Using the right hand rule for positive current flow, the thumb represents the direction of the positive current flow, the forefinger represents the magnetic field and the index finger represents the direction of the resulting force.

In the upper path, the force is downward. For the lower path, the force is upward. If we use an electron, the picture changes but the results are the same. If the present is larger than the past a positive or negative charge will experience a slight force of gravity pointing at the center of the motion. An electron will have a slight gravitational force pointing toward the proton within a hydrogen atom.

If we look at all the atoms within the Earth, there will be a gravitational force vector from every electron toward the center of their respective atoms. Within the protons and electrons, internal AC oscillations of dot current flows will produce additional gravitational force vectors pointing toward the center of the various particles. The same is true of pure planar DC dot current flows for the magnetic moments, etc. They will form a distribution in all directions and also contribute to the net gravitational field.

At a far distance we can look at all the vectors and draw force lines far from the Earth. We can then find the gravitational center of mass of the Earth. All the vectors will tend to have a component, which points toward the center of mass of the Earth. When we add all the vectors up we obtain the Earth's gravitational field

The gravitational field of the Earth is the vector sum of all the individual fields of each atom, proton, neutron, and electron. The field is caused by the net force of a moving charge in orbital motion around a center. As the electrons, protons, neutrons, atoms, and the Earth expand; there is a net electrical force of each piece of mass within the Earth. The vector sum yields a center of mass and a center point of the gravitational field of the Earth.

The same is true of the moon. The attraction of the Earth and moon can be viewed as a net force due to the individual vectors. This can be viewed as similar to photons from afar moving in all directions toward the Earth and moon. Thus:

>>>>>>> Earth .... Moon <<<<<<<   FIGURE 12-2

In Figure 12-2 we see photons from the left hitting the Earth. These do not hit the moon. There are also photons hitting the Earth in every plane. The same is true of the moon. Between the Earth and moon no photons appear. The gravitational field between Earth and moon acts as if external forces pushed them together. In Figure 12-2, the photons push the Earth and moon together.

The gravitational field is independent of electrical polarity. As long as a charged particle orbits a center, it will produce a gravitational force due to expansion. The force is the net force between a charged particle and the memory of the magnetic field of that particle. Gravity and also mass are caused by the interaction between the electrostatic field of a particle and the space-time memory of its magnetic field.

Spherical orbital motion is required for mass in all directions. An electron has mass in all directions when it cycles back upon itself in all directions. A photon will not have mass in the direction of travel when moving at light speed because it does not have forward to rearward repetitive cycles. It will have mass in the perpendicular direction since patterns of dots can form screw type motions or linear motion with perpendicular circular motion. Forward dots will produce patterns which rearward dots encounter. This will remain in space-time memory for awhile as the dots pass. This produces perpendicular mass for the photon. Photons do have inertial mass in the forward direction however. Thus:

Gravitational Mass = Spherical type motion of charged particles   (12-3)

In equation 12-3, we need a spherical magnetic field for mass to exist. An electron moving within itself or moving in the Bohr orbit produces a spherical magnetic field, which exists in space-time memory. The magnetic field also exists within a decaying oscillation. We can fast-forward the image of the electron within the Bohr orbit. The pattern of the electron exists all over the orbit with internal standing wave patterns and with external motion.

Once we expand the electron within the image of the electron or beyond the null point of the magnetic field, we produce a gravitational force. If the electron did not expand, no gravitational force would exist. If the universe stopped expanding, the Earth would blow up. The same is true of the protons and electrons. Someday the protons and electrons will be destroyed when the universe almost reaches maximum expansion. The final explosion will enlarge the universe a little more, and then the contraction will begin. Let us now look at a mass-less photon.

No Gravitational Mass = Linear or non-cyclical motion in direction of travel (12-4)

In equation 12-4, a photon has no mass in the direction of travel when the charged particles move in a spiral pattern or screw pattern, or straight line pattern in the direction of motion. If we cannot produce a steady state magnetic field, which repeats itself in standing wave or moving standing wave patterns, then we have a mass-less photon in the direction of travel. The photon has gravitational mass perpendicular to the direction of travel. Patterns of dots can phase-lock in screw patterns so that rear groups of dots phase lock with the space-time memory of front groups of dots. This will give them gravitational mass in the Y-axis and Z-axis.

The dots will produce photons, electrons, neutrons, and or protons but mass requires a steady state type configuration, which will enable the past and present and future to coexist. The photon moves like the thread of a screw at the speed of light, which does not enable a magnetic image to occur in the direction of travel. If you slow it down, it will develop mass as previously discussed. There will be a loss of inertial mass and a gain of gravitational mass in the direction of travel as the photon moves slowly in various media or heavy gravitational fields.

Everything is made up of dots. From Equation 11-16 we see that mass is not a real individual quantity. Mass is merely a property of the electromagnetic field. Let us look at Figure 12-1 again. If we push upward on the charge Q, it will move upward at the top of the figure and produce a downward force according to the right hand rule. When we look at the orbit on the lower portion of the figure, the charge Q will be moving in the dotted portion of the magnetic field. This will not produce an upward force but a downward force. We get two downward forces when we push upon the orbiting charge Q. When the charge moves upward on the right side and downward on the left side, there is no net force.

The result is that applying a force to a single hydrogen atom will produce two forces per orbit, which opposes the applied force. Of course the electron will move in all orbital directions and we have to add the absolute value of the sine of the angle but in all cases over the entire orbit, the forces will oppose the applied force.

If we take a block of material with billions of atoms and push it, the electrons and the magnetic field of the electrons will always produce opposing forces. The same is true of the protons. The proton currents are much higher than the electron currents. The proton inner magnetic fields are much higher. The net result is that when we push a mass we displace the electron or proton from its magnetic field.

We say that something has mass. A one ton piece of iron is difficult to push upon a smooth surface. As we push it, every electron is displaced from its image and hits us with a force against us. The same is true for every proton. The neutron does the same thing. We do not fight the mass of the iron. We merely fight billions of electrons and protons with their inner fields and outer fields.

If we push the block of iron in outer space, the counter forces fight us. Eventually we reach a steady state. The object will then continue to move without any applied force. We can argue that only during acceleration that the forces between the electron and its magnetic image exist. The object does not know what velocity it is traveling at. This would be an Einsteinian solution.

We can also argue that a permanent displacement occurs between the electron and its magnetic image. A permanent force exists within the block or iron. We can then argue that a permanent counter-force exists within the iron such as the Doppler masses, which set up permanent counter-forces. In this case a block of iron moving at one hundred miles an hour is different from a block of iron traveling at one million miles an hour. Objects of different speeds are different.

In GG/MM/Doppler space-time, objects at different speeds do have different internal forces. Therefore velocity by itself does produce internal forces. The expansion of the universe at constant velocity will produce constant forces. It can be argued that there is acceleration within the universe. That certainly would produce forces. However, constant velocity will displace an object and its image. The removal of an applied force will certainly reduce the distance from an object and its image but not necessarily eliminate it.

In any event there is no simple acceleration term in Equation 12-2. Of course in standard electrical theory:

$$F = Q V B \tag{12-5}$$

In Equation 12-5 we have a moving charge in a magnetic field which obeys the right hand rule and has a force pointing toward an orbital center. Of course orbital motion has acceleration terms. The expansion of the universe with linear velocity still can produce orbital acceleration terms.

Equation 12-2 can then be viewed in terms of equation 12-5 and Figure 12-1. The velocity of the electron in the Bohr Orbit is:

$$V_B = C / 137.036 \tag{12-6}$$

The flux density B at the distance R is:

$$B = 8 \pi U_o V_B^* / R^2 \tag{12-7}$$

Equations 12-6 and 12-7 can be interpreted in terms of the gravitational force of the singular Bohr atom where:

$$R = R_B \tag{12-8}$$

In equation 12-2 we see that the electron has a tiny force of gravity acting toward the proton which is due to the Bohr velocity $V_B = C/137.036$, and the small flux density difference shown in Equation 12-7. This small difference is caused by the displacement of one Bohr orbit from another due to the expansion of the Bohr Orbit at the velocity V_B^*. Thus:

$$B = 8\pi U_O V_B^* / R_B^2 \tag{12-9}$$

Equation 12-9 shows the gravitational force in the Bohr Orbit. Looking back at Equation 12-7 we can say that the electron in a second atom encounters the magnetic field of the first atom at a distance R with the resulting gravitational force between the two.

If the Bohr orbit did not move, all these forces would be zero. The little tiny velocity produces a tiny differential flux density, which produces the gravitational force. It can be argued that space itself is the field. The electrical forces merely compress space as per Einsteinian concepts. The gravitational force is merely the ether's response to the electrical compressive forces caused by the expansion of space.

Of course it can be counter-argued that space and the field are one and the same. This was discussed in Chapter 9. We can always look at the electrical field as waves and we can replace the charge Q with radians per second. This will give us a wave interpretation of things. It is an alternate way of looking at the universe in terms of radians per second, meters, and seconds.

The Earth's gravitational field is easy to understand as a total vector sum of all the little gravitational vectors of all the electrons, protons, and neutrons within the Earth. The same is true of any other body. The attraction of the Earth and moon can be viewed by looking at force vectors at large distances and seeing that Earth and moon will be pushed together. This is especially true when we look at the washing machine effect of the billions of atoms involved.

Let us return to the two lead balls in the lab. They are attracted to each other. It is easy to understand why each tends to be attracted to the Earth. By looking at vectors far from each ball, it can be said that the electromagnetic field of the universe pushes them together.

However, one could also argue that the forces described in this section could be added vectorially to produce the Earth's gravitational field but are not the main mechanism for the force between Earth and moon and the two lead balls in the lab. Thus we need another law of universal gravitation which describes the force between two external objects such as the two lead balls in the lab.

We know that a very complex and powerful electromagnetic field exists everywhere. We know that the individual vectors of the atoms of the Earth produce a single field centered at the center of gravity of the Earth. Let us now look at a law of relativity. In the figure below, we have two observers and a mass in the middle.

[Observer A]   [Mass]>>]

              [<< [Mass]   [Observer B]

FIGURE 12-3

In Figure 12-3 we have observer [A] on the left looking at the mass. As the universe expands, the mass moves to the right while the memory or image of the mass remains. This produces a gravitational force field toward Observer [A]. Therefore, every electron, proton, and neutron in the mass has a force field which tends to push the mass toward Observer [A].

Since the expansion of the universe is equal everywhere, Observer [B] does not see the mass moving toward it. Observer (B) sees it moving away. Observer [B] also experiences a force field of gravity caused by the motion of the mass with respect to its image or memory of its past.

Both Observer [A] and Observer [B] also move away from each other. They both experience universal gravitation due to the motion of the mass and its image. Yet Observer [A] sees the object moving away from its image and so does Observer [B] but to the ordinary mind, they both think the other observer must be wrong.

This is similar to the Einsteinian clock paradox. Physically in ordinary three dimensions, it is impossible for a object to move to the right with respect to its image, when viewed by an observer on the left; and for the same object to move to the left with respect to the same image, when viewed by an observer on the right. Yet, at the very tiny atomic dimensions involved a mirror image effect exists. Thus we have the law of images:

LAW OF IMAGES: UNIVERSAL GRAVITATION DEPENDS UPON TWO OBSERVERS WHO SEE THE SAME THING DIFFERENTLY AND AS MIRROR IMAGES.

We can now write a law of space-time universal gravitation.

LAW OF UNIVERSAL GRAVITATION: An object will always appear displaced backwards from its memory or image when viewed by any observer in any direction. A gravitational force will exist which pushes the object toward the observer. The force is due to the force between the object and its image in the direction of the observer.

The two lead balls in the laboratory are a more simple case. Each has a present and future, which is constantly expanding. There is a constant force from each ball in all directions. However the forces in all directions have little meaning unless they line up to another object. The two lead balls line up and are pushed toward each other. All the forces involved occur at the atomic dimensions.

The Earth sees a moon, which is expanding and moving away from its image. If we look at any one electron on the moon or any current loop, we will find that the electron is moving away from its image and a force exists pointing toward the Earth. Every electron, proton, and neutron within the moon has a force vector pointing at the Earth.

Within the moon, all the vectors add up to the center of gravity of the moon. External to the moon, all the current loops and spheres of current flow produce force vectors all over the universe but this Earth is the closest mass and enjoys the first priority of the moon's gravitational vector.

We see that internal to an object, the gravitational forces tend to find a common center. Once we are external to an object, the gravitational force vectors due to the expansion of the universe tend to appear different to different observers. Opposite observers tend to see the mirror image. However, we live in an electrical universe and not a mechanical one. Electrical theory governs the laws of physics. In electrical theory, two observers can see the mirror image of things.

With this section, it is now clear why the two lead balls attract each other. The concept leads to the two observers who see things as mirror images. Surely this is a problem similar to Einstein's clock paradox. Our minds find it very difficult to visualize such things. Yet, it appears to be the way gravity works.

Man will always find things to perplex him. We now have mirror image gravitational relativity. Thus within the gravitational field itself we have a mirror image effect. Surely it is an amazing and perplexing universe.

SECTION 12-2: GRAVITATIONAL RULES

In this section let us look at the rules of gravity. The rules are as follows:

GRAVITATIONAL RULE 1: The force of gravity is an electro-magnetic force of attraction between the universe of today and the universe of yesterday where today, the present, and yesterday, the past, are only milli-seconds apart.

The force of gravity is caused by the differences between the electromagnetic field of this present split milli-second and the fields as of a split milli-second ago. The second rule is:

GRAVITATIONAL RULE 2: The force of gravity is due to the expansion of the universe and the memory of yesterday.

If we take a simple current loop and expand it slowly, there will be a positive force of gravity between the current loop of today and the current loop of yesterday, which remains in the memory of space and time. A force will exist that attempts to prevent the expansion of the current loop.

GRAVITATIONAL RULE 3: The expansion of the universe is caused by the decompression of the various compounded charges Q within particles and dot charges Q_D with outer-charge Q. As the charges decrease, the universe expands.

As the compressed universe decompresses, space becomes uniform and everything decays. The stars and planets eventually explode. Eventually the electrons and protons also explode. All along the way from big bang toward maximum expansion the universe discharges. We can look at the charge +Q as a mountain peak and the charge -Q as a deep valley. As the mountain's peaks and valley's decay, we obtain a flat smooth surface. The universe operates in the same way. When it expands toward infinity, it becomes uniform with no differences between plus and minus. It becomes pure empty space.

GRAVITATIONAL RULE 4: As the universe expands gravity appears as an attractive force between nearby clumps of matter. As the universe contracts, gravity appears as a repulsive force between fundamental dots.

The expanding universe permits stars and planets to form as the gravitational force external to clumps of matter tend to push them together. This is the universe, which we know. When the universe is completely destroyed at full expansion or little bang, volumes of dots will compress but the forces of gravity will be negative and will prevent the formation of anything.

Electrons, protons, neutrons, planets, and stars all require a compression of the electromagnetic universe toward multiple pinpoints. Of course the big bang is the major producer of things. However, in empty space unbalanced gravitational forces could produce galaxies at any time. There are plenty of non-linearities in the universe and compression can occur. The big bang would do the major job but solar systems, and galaxies could be forming today in non-linear areas of space-time.

GRAVITATIONAL RULE 5: The gravitational field of this Earth and other bodies is due to the slow expansion of the electrons, protons, neutrons, atoms, molecules, and atomic spacing. This produces a constantly expanding Earth, which is always larger than the Earth of yesterday.

Everything contributes to the gravitational field of the Earth. Every electron, proton, neutron, and atom contributes. The gravitational field at any point on the Earth or surface of the Earth is the vector sum of all the individual contributions of every charge and moving charge within the Earth.

GRAVITATIONAL RULE 6: Every electron or proton has an internal gravitational field pointing toward its own center. This gives the particle a self-gravitational force and the property of mass.

The electron by itself in free space expands and has a spherical force toward its own center. This gives the electron both mass and inertia. The same is true of the proton. These fields are caused by AC oscillations of groups of dots.

GRAVITATIONAL RULE 7: The Universe is purely an electromagnetic field and the property of mass is a property of the field. In particular, the property of mass is due to the electromagnetic forces acting upon an electron or proton as it expands spherically.

The electron has an overall charge Q and current flow within it. The internal magnetic field of the electron has a past component and a present component. When the electron expands, there will be compressive gravitational forces acting to stop the expansion. When the electron moves there will be current loop attractive forces between the electron of yesterday and the electron of today. This will act to prevent the forward motion.

In free space, if you push on an electron, the forces between the electron and its image (or shadow) will act to prevent the motion. This will cause the counter-force of inertia. As the velocity increases and the applied force decreases, there will be a reduction in the distance between the electron and its image. The maximum time or distance between the electron and its image occurs when the applied force is a maximum. Later counter forces are built up and we get a phase lag between the electron and its image. This depends upon the velocity.

GRAVITATION RULE 8: The distance between an electron, or proton and its image increases with velocity.

These distances are very tiny for normal velocities. At one hundred miles an hour the distance between an electron and its image will almost be zero. At one thousand miles an hour, the distance between an electron and its image will be much more. When we move toward the speed of light in a cyclotron, the electron and its Doppler image will become a line and become quite long. The image will no longer be confined to nuclear dimensions. At low speeds the image will be 180 degrees away. Near the speed of light it will lock on the electron.

GRAVITATIONAL RULE 9: The force between two masses depends upon the amount of particles or mass in mass 1 and the strength of the gravitational field of mass 2 and visa versa.

If you take a mass near the Earth, there will be a force between the mass and the Earth. The force acting upon the mass will depend upon the amount of matter in the mass, and the strength of the gravitational field of the Earth. Likewise the force acting upon the Earth depends upon the amount of matter within the Earth and the strength of the gravitational field of the mass.

This is standard basic physics, but it points out that a mass one thousand miles from the Earth has a much stronger force acting upon it than the same mass one thousand miles from the moon.

This implies that the force of gravity depends not only upon the particular hydrogen atom's self-gravity and expansion and the expansion of the universe but also upon the intensity of nearby matter. Since only the electromagnetic field exists and since all we have are charges and current flows, an electromagnetic gravitational field from the Earth must exist right at the location of the mass itself.

The gravitational field is not a field at a distance but actually exists everywhere. Thus:

GRAVITATIONAL RULE 10: Every electron and proton within the Earth has electrostatic and magnetic fields everywhere in the universe. When you move away from the Earth and look within space you will find the field and motion of every particle within the Earth. If you take a piece of space the size of a neutron, you will find the image of billion of billions of tiny moving charges within it from external objects.

The tiny piece of space will be electrically neutral. Yet if we subdivide it we will find the equivalent of tiny voltages and currents from all the atoms of the Earth. Every piece of space is electrically alive and every piece of space tends to have attractive and repulsive forces acting within it. As the charges decay, the universe expands but at the same time every piece of space will have a net attraction to every other piece of space.

When a single atom stands away from the Earth, the electron and the proton will react with the Earth's electromagnetic gravitational field. Near the Earth the density of the total field is large and the atom will have a strong gravitational force acting upon it. When we are far from the Earth, the density of the total field is small and the atom will have a weak gravitational force acting upon it.

GRAVITATIONAL RULE 11: The center of gravity of the Earth is located at the point where the electrostatic fields and magnetic fields of every atom within the Earth balance out.

The Sun and Moon shifts the net gravitational center but for the Earth itself, the center of gravity is a point of electromagnetic balance. The forces of every atom upon the Earth and the atmosphere as well, depend upon the inverse square law and have a single null point. Every other point within the Earth has a force vector which points to this electrically neutral point.

When we move far from the Earth, the field weakens but every point still points toward the electrically neutral center point of the Earth. Every point in the universe sees the center of the Earth as one vector component of its gravitational field.

GRAVITATIONAL RULE 12: The gravitational force acting upon an object is the vector sum of all the forces from all the centers of gravity of all the objects.

If you have an area of free space with five masses, each mass will have a force of gravity dependent upon its own mass and the four other gravitational fields of the four other objects. Within linear levels of gravity, the various forces will appear similar to partial pressures from different gasses as per the general gas law. Each force only depends upon the mass of the object and the other masses with their separate fields. Of course, this depends upon linear levels of mass. Once we go to super-dense masses, the effect of one light mass upon another will be negligible as compared to the effect of the huge non-linear masses.

The Earth is subject to the force of the Sun, the moon, and all the other planets. Comets and asteroids will have temporary effects. The entire galaxy will have effects as well but huge distances reduce the effects of far objects due to the inverse square law of distances.

GRAVITATIONAL RULE 13: An object in free space moving with velocity V in a straight line far from any gravitational field will move in a straight line while slowing at a cosmic rate.

Standard classical physics states that an object will move in a straight line forever. However, there is nowhere in the universe where you can travel forever before encountering first a weak gravitational field, and later a strong gravitational field. In addition, just as the photons experience a red shift and the protons experience a red shift, all objects will lose their kinetic energy at a cosmic rate and come to a halt eventually.

GRAVITATIONAL RULE 14: An object moving in a straight line in free space which encounters a gravitational field of a galaxy or star, and which moves tangential to the gravitational field will slow in the direction of motion. The force of slowing depends upon the speed of the object, its mass, the strength, and the direction of the gravitational field.

In general, the gravitational field exists everywhere. Between galaxies it can be very weak. We can have areas of space where the net gravitational force is zero. This can occur when the masses are equal in all directions relative to the inverse square distance law. We can have a moving object with zero net gravitational force, which encounters a star in a tangential orbit.

When the object moves directly toward the star, it will gain velocity. When the object moves tangential to the star it will slow due to the interaction with the gravitational field of the star. The reason for this is that the gravitational field will induce perpendicular forces in the object. This causes vector forces opposite to the direction of motion of the object.

When the object has more mass, the forces will be greater. When the object has a greater velocity, it will have less gravitational mass in the forward direction and more inertia in the perpendicular direction. The ability to turn the object will decrease with speed. When we reach near the speed of light, the inertial shape of the object changes from a sphere to a tube, which moves ahead of the gravitational center. Electrons moving near the speed of light can pass a star without much slowing or turning.

If the object was moving slow such as our Earth, and encountered a star, it would slow in the direction of travel. If we took a set of axis for the object, it becomes clear that if the object is traveling only in the minus X direction, it will gain speed in the minus Y direction and lose speed in the minus X direction. The gravitational force on the object always slows the object in the direction of travel as per the simple electrical forces. Let us now look at the perpendicular direction.

GRAVITATIONAL RULE 15: An object moving in free space tangential to a gravitational field will find a force of gravity pointing toward the center of the gravitational field. The velocity of the object will increase in the direction of the field.

The Earth always has a force of gravity pointing toward the center of the sun. The Earth will always increase its velocity in the direction of the gravitational force. In classical physics, we look at the gravitational force between the sun and the Earth as:

$$F_G = G M_E M_S / R^2 \qquad (12\text{-}10)$$

Equation 12-10 states that there is a gravitational force between the Earth and Sun equal to the gravitational constant G times the mass of the Earth M_E, times the mass of the Sun Ms, and divided by the distance squared. The counter-force is due to the centrifugal force. Thus:

$$F = M_E V^2 / R \qquad (12\text{-}11)$$

Equations 12-10 and 12-11 are mechanical ways of looking at the forces between Earth and Sun. Yet, in electrical theory, these forces do not exist. For an object traveling in the counterclockwise direction, the electrical gravitational forces would be:

$$F_{GY} = -G M_E M_S \sin(a) / R^2 \qquad (12\text{-}12)$$

$$F_{GX} = -G M_E M_S \cos(a) / R^2 \qquad (12\text{-}13)$$

The vector sum of the forces is:

$$F_G = G M_E M_S / R^2 \qquad (12\text{-}14)$$

Equation 12-14 is identical with equation 12-11. However, if we place the Earth at the top of a circle and spin it around the sun in the counterclockwise direction, the forces in the Y direction will increase the velocity in the direction toward the sun. At the same time the X velocity or tangential velocity will decrease.

Once we move 90 degrees or at the -Y axis, the Y velocity will be maximum and the X velocity will be zero.

The electrical way of looking at the Earth's orbit produces two gravitational forces which are perpendicular to each other and which produce a circular orbit. The centrifugal force appears to be an imperfect concept. In reality the centrifugal force is the vector sum of two perpendicular electrical forces. A moving object generates both horizontal and vertical forces with respect to an arbitrary axis of the gravitational field.

When a moving object exists within a gravitational field, both axial and perpendicular forces are produced. In free space beyond a gravitational field, there is neither axial nor perpendicular forces produced. There is only the self-induced forces of the inertia of the mass itself. Let us now look at the center of gravity of two equal objects.

GRAVITATIONAL RULE 16: The electrical center of two equal objects will be exactly halfway between them at the point where the fields of all the electrons and protons exist in perfect null. The field within the center of each object will be shifted toward the center of the combined field. As we move outward away from the two objects, the two combined fields will no longer appear as two fields. They will tend to appear as a single field. This causes a gravitational force, which pushes the objects together.

The gravitational field is similar to a magnetic field. However, it is a combined electrostatic field and magnetic field. It tends to be electrically neutral on the large scale. At the microscopic level we have a very fine E/M field. This field is attractive to itself. The field around two objects tends to bring them together.

If we look at the field around the two objects and then enlarge it in all directions slightly, this will produce the field of yesterday and the field of today. Then we see that a force exists that presses the entire package together.

The fields by themselves produce patterns, which lock together two objects.

The force between the objects which tend to make them move together are caused by the general expansion of the universe which in turn is caused by the decay of the charge Q over time. Let us now look at the distance between two objects over time.

GRAVITATIONAL RULE 17: Two objects in orbit in free space will have a force of gravity between them. As the present expands compared to yesterday, the standard ruler of the universe will expand. The gravitational force between the objects will tend to negate the general expansion of the universe. If the objects only gain in motion toward each other as much as the expansion, the objects will remain a constant distance apart as measured by a non-varying ruler, but the objects will constantly move closer to each other as measured by a ruler which varies with the expansion.

The universe can expand and the ruler can expand. The force of gravity will act to maintain the general distances for orbital motion. The net result is that the universe can expand while the galaxies can appear to get closer together.

Rule 17 is just one factor in the motion of galaxies. There is initial kinetic energy, the loss of rotational energy, etc. Also, two objects in a straight line will not maintain their distance, they will come together.

The two lead balls in the lab are constrained. The universe is getting bigger and there is a force between them. Yet, they are not free to move together except for a short distance. One could argue that the force of gravity between them is caused by their effort to maintain their original relationships. One could argue that each ball attempts to maintain its own inertial balance while the distance between them keeps increasing as the universe expands.

We see that the gravitational field is really the electromagnetic field. People look at the hydrogen atom and declare it electrically neutral. Yet, at small distances, the electrical fields of atoms react to form binding energies of atoms and chemical bonds. When we add many protons together we get defined vector angles between the protons. When we look at the entire Earth, we see so many atoms that we view everything as electrically neutral. Certainly we do not find points in space with huge charges coming out of nothing.

In spite of this we either get neutral space consisting of equal amounts of plus and minus voltages or we break down space into very tiny dot voltages of plus and minus nature. The net result is that space appears electrically neutral while in reality it carries extremely large amounts of charges and currents.

We like to think that things form standing waves and that is the end of the electrical nature of matter. Yet, standing waves are only that points in space from an atom or group of atoms where things appear quite neutral beyond those points.

In reality space itself is a living electromagnetic field. Although an electron hits the screen of a monitor display, its effect reaches far and wide across the universe. It can be argued that eventually its effect is too small to count and that we can quantize the electric field of a single electron to some finite distance, such as a million miles. The quantization of the resulting field a billion miles away may require one billion electrons to do the same job as one electron a mile away. All the electrons in a small ball may be necessary to be felt at the radius of the universe.

The gravitational field is a very simple field. It is standard electrical theory except for the addition of space-time memory. Of course if we look at the capacitance and inductance of space, we can define space-time memory as equivalent to the electrical delays of L/C electrical circuits. Yet, it is merely a different way of looking at the same thing. The universe does have space-time memory, which produces mass, inertia, and gravity. Without space-time memory, the universe would not hold together.

Let us now look at two objects of equal size and mass in free space. The following gravitational rule applies:

GRAVITATIONAL RULE 18: The gravitational field of two equal masses in free space has a center halfway between the masses at the intersection of tangential lines drawn between opposite sides of the masses. As we move outward from the masses to planes of equi-potential, the gravitational field will be egg shaped or an ellipsoid. As we move further away, the field will turn into a perfect sphere. As the universe expands the gravitational force will be applied from the sphere toward the common center of gravity. The net result is that the spherical forces will crush the elliptical forces in an attempt to produce a sphere everywhere. Therefore the two masses will be pushed together.

We see that two objects form a common center. We see that far from the two objects, the combined field is a perfect sphere. If we look at the makeup of this field we find that each electron within the masses produces a spherical pattern moving far into space. The motion of the electron produces a sphere within a sphere. When we add the proton we attain another sphere within a sphere and a complex pattern of voltages at far distances from the object. As soon as we add billions of billions of electrons and protons, we attain an extremely complex pattern of fields. In general, the vector sum of any piece of space tends to be neutral. However this neutrality is actually the result of equal and opposite voltages from a huge number of atoms.

In general, we could say that the resulting field looks like random noise of varying intensity. The electrical noise is very strong at the common center and weak far away. Since gravity depends upon the electrical pattern of a split second ago as compared to the electrical pattern of the here and now, gravity is independent of whether the signals are clean and pure or merely complex noise patterns. Thus:

GRAVITATIONAL RULE 19: The force of gravity everywhere in space only depends upon the total plus and minus field strength and the expansion of the universe. Charges moving in random order within an object produce complex reflections far away and produce a gravitational field, which is independent of whether the charges move in simple patterns or in random noise patterns.

Since the gravitational force compares the universe of today and the universe of a split second ago, gravity works quite well within chaos. If we have an explosion and a solar system or galaxy exists in pure chaos, a center of gravity of the mess will still exist.

Ordinary electrical circuits work with meaningful signals except for some noise machines for assistance in sleeping and testing. The gravitational forces exist both in chaos and in harmony. The heat of the sun produces chaos. Yet, the gravitational field of the sun doesn't care if the atoms of the sun have any relationship to each other or not.

Upon this Earth we have mixtures of harmony and chaos. The general gas law could be considered properties of chaos. If you look at air molecules far from the

Earth, you will not get nice patterns of the electrical fields. You will get chaos. The gravitational field does not care. It is only interested in the intensity of the voltages and currents from each source, and the amount of sources.

We see that gravity is the most basic and simple force. It works in harmony and it works in chaos. Gravity is a very simple property of space and time.

Let us now look at the property of inertia as we move from free space to a strong gravitational field.

GRAVITATIONAL RULE 20: The force necessary to bring an object up to a particular velocity in pure free space is less than that required to bring an object up to speed in a moderate gravitational field. This is much less than required to bring an object up to the same speed in a strong gravitational field.

Just as the gravitational force upon an object depends upon the external gravitational force, the same is true of the force necessary to move an object. When we move an object in pure free space all we have is the self-gravitational field of the object.

When we move an object within a strong gravitational field, the moving object inter-reacts with the gravitational field to produce eddy currents. The stronger the gravitational field the stronger the currents, which produce heat and slow the object.

Spacecraft traveling at very high speed in pure free space must avoid strong gravitational fields. If they hit a strong gravitational field, they will slow greatly and at the same time, they could burn up. Future space travelers must insure that when traveling at near light speed the strong fields of planets and local stars are avoided.

GRAVITATIONAL RULE 21: Every cubic meter of space in the universe attracts itself and every other cubic meter of space in the universe.

The electromagnetic fields exist everywhere in the universe. Every cubic meter of space contains the projected image of billions of billions of current flows. These projected images cause compression of space due to the attraction of the field of yesterday with the field of today.

This means is that gravity holds together every cubic meter of the universe. Even without any mass within that cubic meter, the force vectors still exist. In this manner, empty space still contains a gravitational field, which tend to compress the empty space.

The compression of pure empty space enables all the gravitational fields within the universe to be interconnected.

SECTION 12-3: DOPPLER GRAVITY

In this section let us look at the gravitational force from a Doppler point of view. Equation 12-2 provides us with an electrical equation for the gravitational field caused by a charge, which moves in a spherical orbit. The result is a force toward the center of rotation of the charge.

That is one way of looking at gravity. It is an electrical way. One could use the same equation in terms of waves. We could also look at it from a mechanical viewpoint. A mass revolving around a center point experiences force vectors toward the center of rotation due to expansion of the distances.

The spherical Doppler equations can be written in electrical terms or mechanical terms. First let us look at a moving Doppler mass.

$$Mx_R = Mo\,[1-(V/C)^2]^{1/2}\,[C/(C-V)] \tag{12-15}$$

$$Mx_L = Mo\,[1-(V/C)^2]^{1/2}\,[C/(C+V)] \tag{12-16}$$

In equation 12-15 we see that the mass moving to the right has a decreased gravitational mass but an overall increased inertial mass due to the Doppler. It also has a projection of Doppler length and Doppler inertial mass far to the right as we near the speed of light. If we go fast enough, the projection reaches the radius of the universe.

The mass to the left decreases gravitationally and also decreases due to the Doppler. Now let us look at the same mass expanding spherically.

$$M_A = Mo'\,[1-(V/C)^2]^{1/2}\,][C/(C-V)] \tag{12-17}$$

$$M_B = Mo'\,[1-(V/C)^2]^{1/2}\,[C/(C+V)] \tag{12-18}$$

In equations 12-17 and 12-18 we have two Doppler masses. In Equation 12-17, the mass M_A is very large since as we move toward the speed of light, the second term is much larger than the first term, which moves toward zero. In equation 12-18, the mass M_B is small because the first term is small as the velocity reaches light speed and the second term is equal to one half, (1/2).

We can say that M_A is the mass of the entire universe. We can say that M_B is the mass of the proton or the electron. We can then solve for the ratio just as we did with charge. The important point is that the proton, neutron, or electron has the entire universe pressing upon it due to the Doppler. The above equations become the Doppler force equations.

The entire universe is expanding outward. The entire mass of the universe is expanding outward. The Doppler equations translate that into back-pressure acting upon the entire universe. We can look at it from an electrical perspective. The Bohr atom is expanding and electrical pressures come to bear. Likewise we can look at it from a wave or mechanical method. The Doppler frequency due to expansion produces a back-pressure everywhere.

Electrical answers can be replaced by mechanical wave answers. The electrical attraction between the Bohr orbit and the memory of itself become the expanding standing wave, which experiences a Doppler back-pressure.

We have different ways of looking at the same thing. We can always say that charge is equivalent to radians per second. We can convert an electrical universe into a mechanical universe. We will get the same answers but sometimes some problems are easily understood in terms of waves and other problems are better understood in terms of electrical theory. In any event we live in a Doppler universe.

CHAPTER 13: THE DUAL DOPPLER EARTH

SECTION 13-0 INTRODUCTION

In this small chapter we will expand upon the dual Doppler Earth as discovered in the study of orbital space-time. A dual Doppler electron exists in the cyclotron opposite the original location. This is the point where the Doppler frequencies merge. At lower speeds, it is exactly opposite the electron. At higher speeds, it is closer to the electron. The location will be specified in this chapter.

Every cyclical object in the universe has a dual Doppler image. There is an image of this Earth and there is an image of us. Let us now investigate the properties of the Doppler image.

SECTION 13-1 THE DOPPLER IMAGE

Let us look at the electron moving in the cyclotron and calculate the location of its Doppler image. The image is the point where the number of cycles to the right matches the number of cycles to the left. Thus:

$$f_R = f_{Og} [C/(C-V)] \qquad (13\text{-}1)$$

$$f_L = f_{Og} [C/(C+V)] \qquad (13\text{-}2)$$

$$f_{Og} = f_O [1-(V/C)^2]^{1/2} \qquad (13\text{-}3)$$

In Equation 13-3 we have our Einsteinian orbital clock frequency reduction formula. This is caused by the build up of steady state Doppler energy throughout the orbit. In addition the junction point of the Doppler waves produce an image. This occurs in two dimensions primarily and with corresponding images in the third dimension. A match occurs which fills in the third dimension. The electron then has an image located where the number of cycles of f_R match the number of cycles of f_L. The image point in degrees will be:

$$\theta = 180 (C-V) / (C+V) \qquad (13\text{-}4)$$

In equation 13-4 we find that the image angle θ is 180 degrees when the electron is moving slowly. When the electron is moving close to the speed of light C, the image angle θ is 0 degrees. The image at close to the speed of light is right in front of the electron. Of course at that speed the electron is a line of energy and the simple linear equations need to be replaced.

At the speed of light, the Doppler dual image for the electron appears to be at its own location. An electron tends to double its gravitational energy at its location when traveling at the speed of light. Of course the entire pathway is also filled with orbital gravitational energy and multiple images can occur as we pump more and more energy into the system.

In effect, we can always double the energy of anything when we move toward the speed of light in linear motion or by compressing a star, and or compressing the entire universe. The Doppler image duplicates the object when the speed approaches light-speed. Likewise if you compress the universe enough, you get neutrons which have twice the energy level of ordinary neutrons. The calculation for the minimum radius of the universe was based on the size of the neutron. However, it could go to half the volume. Thus:

$$\text{Volume of neutron at big bang} = \tfrac{1}{2} \text{ Volume of neutron} \qquad (13\text{-}5)$$

$$R_{Nbb} = (0.5)^{1/3} R_N = 0.7937 R_N \qquad (13\text{-}6)$$

$$R_{Nbb} = 1.04736\text{E-}15 \qquad (13\text{-}7)$$

The single neutron at big bang would have twice the energy of the neutron in the universe today, and would be reduced in size to approximately 79 percent. This will change the light speed hysteresis loop calculation accordingly.

We see that the image of the neutron merges with the neutron at big bang. In effect we have a universe of physical energy and a universe of soft energy or photon energy. At the big bang we had the merging of the hard energy of the dots and the soft energy of the motion of the dots at light speed. The giant neutron and the ghost of the giant neutron were one at the big bang. We could also call this the soul of the neutron. Body and soul were one at the big bang. Scientifically we could say that:

$$\text{Energy of Body} = \text{Energy of Doppler Image} \qquad (13\text{-}8)$$

In equation 13-8 we see that at the big bang, the Doppler image energy was identical with the material energy. The mass of the universe at big bang was twice the mass of the universe at present. Of course we are riding an e^x curve so that the mass of the universe at big bang approached infinity. However the normalized mass was:

$$M_{Ubb} = 2 M_{Uo} \qquad (13\text{-}9)$$

Let us now look at the Dual Doppler Earth. This Earth is a Doppler image of our Earth. This is similar to a hologram of the Earth at the Doppler Earth location. The speed of the Earth is:

$$V = (2\pi\, 9.3\text{E}7) / [(365) \cdot (24)] = 66{,}700 \text{ miles per hour} \qquad (13\text{-}10)$$

The Earth is traveling at 66,700 miles per hour using the simple 365-day orbit and 93 million miles for the distance to the sun instead of the exact numbers. The accuracy of this calculation is not very significant. Rough numbers are fine. In miles per second we have:

$$V_E = 18.53 \text{ miles per second} \qquad (13\text{-}11)$$

Since the speed of light is 186,000 miles per second, the Doppler angle to the Doppler Earth is:

$$\theta = 180 \text{ degrees} \qquad (13\text{-}12)$$

Since our Earth speed is so low as compared to the speed of light, the amount of shift in the image angle from 180 degrees is only one part in ten thousand. The most important thing is the Doppler energy of the Doppler Earth or the equivalent rest mass. The mass of the Earth is approximately:

$$M_E = 1.4E28 \text{ pounds} \qquad (13\text{-}13)$$

Using this weight for the Earth, let us now calculate the rest mass of the Doppler Dual Earth with foci on the opposite side of the sun from this Earth.

$$M_E' = M_E [1 - [1-(V/C)^2]^{1/2}] / [1-(V/C)^2]^{1/2} \qquad (13\text{-}14)$$

In equation 13-14 we see that the last term is effectively unity at the small Earth speed. We also see that the Doppler Earth is the Earth mass times a small Einsteinian differential mass. Since V is very small as compared to C, equation 13-4 can be reduced to:

$$M_E' = M_E V / 2C \qquad (13\text{-}15)$$

This calculates to be:

$$M_E' = 6.94E19 \text{ pounds} \qquad (13\text{-}16)$$

We see that the dual Doppler Earth contains a huge amount of energy. Even though the Earth is traveling very slowly, the mass of the Earth is so large that the Einsteinian mass increase is significant.

If a spaceship orbited the sun exactly opposite to us at the Doppler Earth foci, the Earth would be visible in all directions like a hologram. It would be a mental image of the Earth in space-time energy.

We exist in the here and now. We also have a Doppler image all over the orbit of the Earth. We have a Doppler weight in proportion to the Earth's Doppler weight. For a 180 pound person:

$$M_{STDI} = 180 \text{ pounds} [6.94E19/1.4E28] = 8.92E\text{-}7 \text{ pound} \qquad (13\text{-}17)$$

Our space-time Doppler Image with foci at the Doppler Earth and distributed through orbital space weighs only 8.9E-7 pounds. It is about one-millionth of a pound. It has very little energy or equivalent rest mass but it is the space-time image of us. It enables us to exist both within our bodies and outside of ourselves.

The meaning of the Dual Doppler Earth and our Space-time Doppler Image necessitates a philosophical study of Doppler space-time. This study begins in Chapter 14. We will look at what we have discovered from a scientific philosophical viewpoint.

PHYSICS & PHILOSOPHY

CHAPTER 14: THE ORIGIN OF THE UNIVERSE

SECTION 14-0: INTRODUCTON

In this chapter we will look at the analysis from the physics section of this book in order to understand how the universe came into existence in the present form. We will look at the origins of the universe from a purely natural physical point of view and also a scientific philosophical point of view. We will primarily be concerned with the single light speed solution. However, we will briefly look at the salient points of the multi-light-speed solution.

SECTION 14-1: IN THE BEGINNING

In the beginning, as far back as we can physically go, the universe existed as a homogeneous electromagnetic field over infinite space and infinite time. It was perfect homogeneity. There was no thought. There was no emotion. There was no intelligence. There was nothing except a homogeneous electromagnetic field of infinite space, infinite time, and infinite potential energy.

Infinite space, infinite time, and infinite potential energy are the main attributes of the neutral homogeneous electromagnetic energy field. There is no magnetism. There is no electricity. There is only a neutral field of pure space without the slightest trace of an imperfection.

If you take a spaceship within this field, you can find no distinguishing features. It is pure blackness. It is pure void. It is absolutely pure nothingness. It is perfect homogeneity. Thus in the beginning we have nothing at all.

Nothingness can be considered as a null point between two opposing forces. Nothingness can also be considered a state of perfect existence. In the beginning was perfection. There was no pain and suffering. There was no man. There was no intelligence. None of that was necessary. All that was necessary was pure nothingness, pure homogeneous space-time.

It did not take one ounce of brainpower to produce the present universe. Not even the slightest thought was involved. All it took was infinite time and infinite space and pure homogeneity.

Space and time exist within a pure electromagnetic field. In the purest sense, it has no kinetic energy. It only has potential energy. Space-time has the ability to be. The space-time bubble that is our universe always existed. We know it existed because we are here. We are part of it. When we look back in time we see that as we move toward minus infinity, the space-time bubble has no features at all.

At minus infinity, both the active energy and the mass are zero. Distance is infinite and time is infinite. The light-speed is also infinite. There is no light but its speed is infinite. The potential energy is huge. Even though nothing distinguishable at all exists, a mathematical space-time entity does exist.

The infinite space-time machine will stay at infinity forever. In the beginning for all eternity nothing existed forever. The space-time machine stands in a relaxed state forever. There is no pain no suffering; no thought. There is no memory. There is nothing at all. There is only a machine. It is a space-time electromagnetic field in a perfectly null state. There is no active energy whatsoever. It is pure absolute death.

Before the universe, pure nothingness existed. It is not a remarkable thing. It is just ordinary math and electrical theory. It is the end point of a spectrum of solutions to the universe, and all the possible universes, which can be produced from the end point. All solutions return to the space-time ether, which permits all possible universes to appear.

We come from a space time machine, which produces a spectrum of solutions of which a state of pure nothingness is one starting solution. It is an inversion machine. The inverse of the state of zero is infinity. Thus the machine produces an infinity of solutions.

The universe we live in is one solution out of an infinity of solutions. We can have multiple light speed solutions. We can have solutions of pure energy and no mass. We can have solutions of pure mass with no external energy. One large lump of matter is a solution. Another solution is standing waves of pure energy. Again it means nothing to us. It has no feelings. It has no life. There are an infinite number of solutions in which life does not appear. The universe can be a beautiful machine or art form without any intelligent life whatsoever.

Singular solutions with singular intelligent life are also possible. Endless universes, one after another are another solution. The space-time machine can produce never-ending possibilities. We are brought into existence by Aristotle's prime mover. It is just a space-time machine, which grinds out man and beast forever.

The universe we live in is just one mode of operation of an electromagnetic field, which spins endless combinations of universes and states of null. It is just a mathematical entity.

In the Dot theory, one mode of operation is a perpetual oscillation from big bang to little bang. This will be explained in the next section.

SECTION 14-2 THE OSCILLATING UNIVERSE

One solution to the universe out of an infinity of possible solutions is that of a universe which oscillates repetitively from the big bang to the little bang. The big bang is a condition where all the mass/energy of the universe is concentrated in a sphere of minimum radius. This explodes as the big bang and the universe expands toward infinity.

The little bang is a condition where all the protons, electrons, neutrons, and photons lose their binding energy and explode as the universe reaches maximum radius. The universe then implodes and starts to compress toward the big bang again. The universe oscillates from big bang to little bang over and over again.

In order to reach a steady state oscillating universe we need to start the cycle at infinity. The electro-magnetic field will compress at the speed of light which is also infinity. It will head toward a single pinpoint. As the universe moves toward a pinpoint, it will gain kinetic energy and mass. It started in a mass-less state at infinite light speed and heads toward a single pinpoint of infinite mass at zero light speed.

The universe moves between the extremes of infinite light speed and zero light speed. It moves between infinite potential energy and zero potential energy. It moves from zero mass toward infinite mass. As the universe moves along this path, a point is reached where space-time splits.

We started with a homogeneous electromagnetic field, which did not have positive and negative characteristics. As we bring the field toward a pinpoint space and time split. We get a space-time hysteresis loop similar to magnetic iron. The path going toward the pinpoint does not match the path moving back toward infinity.

If the path going toward the pinpoint matched the path moving back toward infinity, we would have a simple straight line from zero to infinity. The universe would be a neutral homogeneous blob of nothingness. It would oscillate from zero to infinity and not one thought would ever be produced.

We exist because of a hysteresis loop in space and time. The line from zero to infinity became different from the line from infinity to zero. The split of space-time produced positive and negative. As the universe compressed toward zero, two separate and distinct entities occurred. These were the first two dots of the Dot theory. The electrical charges were huge. The universe was composed of only two dots.

Compression increases electrical charge and expansion reduces electrical charge. At zero radius the two dots have nearly infinite charge whereas at infinite radius the two dots have basically zero charge. The split of space-time and the resulting hysteresis loop prevented the universe from returning to the perfect null state at infinity.

The universe expanded to a maximum radius at very high but not infinite light speed. It then compressed back toward zero size. The hysteresis loop prevented the destruction of the universe at infinity. It also prevented the expansion down to a perfect pinpoint. The net result was that on the next trip, which took nearly infinite time, the two dots became four dots.

The minimum radius enlarged and the maximum radius shrunk. The universe started an oscillation, which increased the number of dots each time. There are now over, ten to the one hundred twenty fifth power, of dots in the universe (10^{125}). The entire universe is made of these dots which are tiny electro-photons. Electro-photons are charged photons. Ordinary photons are neutral and are composed of a balance of plus and minus dots.

Electrons contain an equal number of plus and minus dots with an excess of minus dots. Protons contain more plus and minus dots than the electron with an excess of plus dots equal to the same excess as the electron's negative dots.

As the universe oscillated, the light speed stabilized. We now have a light speed hysteresis loop with an extremely tiny light speed variation over the entire cycle.

The difference of energy within the hysteresis loop is equal to the excess Einsteinian energy of the oscillating universe itself. Although this is a small number, it is energy above and beyond the normal energy. We can call it Einsteinian differential energy. Likewise we can call it spiritual energy. It is the creative energy above and beyond the ordinary. It is the energy of the Mind of the Universe. It is similar to orbital space-time energy but it is large-scale whole body steady state energy.

The universe then reaches a perpetual oscillation. We are on a mathematical operating point of a universe capable of going from zero light speed to infinite light speed. Of course we can have a multi-light-speed solution with coexisting universes but for the present, only a single light speed solution is discussed. We then exist within a space-time machine which started at infinity and which is now finite.

We ride an exponential sinusoid. Time varies with time and distance varies with distance. For a standard clock, which is constantly changing, it appears that the universe takes 1244 billion years to oscillate from cycle to cycle. On a fixed time clock, it really looks like infinite time.

Both the ruler and clock expand each day although the light speed remains constant. The universe expands toward infinity although it doesn't quite look that way due to the changes in both clock and ruler.

The hysteresis loop prevents the universe from being a one shot universe. It also enables space-time memory. Intelligence is only possible when a clock exists. A clock goes around in a circle and repeats itself. An atomic clock is based upon an oscillation or a repeated event. Intelligence requires repeated events. Chaos involves non-repeated events. True random noise does not repeat. Tones, which repeat, and combinations of sounds, which repeat, produce music.

Intelligence requires repetition. The space-time hysteresis loop and the oscillation of the universe produce a situation where repetition occurs. The splitting of space-time into plus and minus dots and the oscillation of the universe produces intelligence. The Mind of the Universe exists because the Universe has a repetitive oscillation. This produces Einsteinian space-time energy, which contains memory.

The universe oscillates and this produces space-time memory. The entire universe starts to become a mind of sorts. Out of pure nothingness and pure homogeneity, a little bit of heterogeneity converts pure nothingness into the start of intelligence and space-time memory.

The electro-magnetic field began to produce a Mind of the Universe. No mind existed at the beginning. It was only the potential energy that existed. Once the universe began to oscillate with a repetitive oscillation, a mind was automatically produced.

The Mind of the Universe is a simple Mind. It is a memory machine. Out of pure nothingness comes a space-time oscillation, which produces memory within the Einsteinian energy all over the universe. Each cycle of the universe increases the number of dots and the heterogeneity potential of the universe. The more dots, the more complex the forms of life. Heterogeneity increases for each cycle of the universe. In addition each cycle of the universe, up until a steady state condition, perfects the memory of the Mind of the Universe. The Mind of the Universe increases in intelligence during each cycle.

The most important increase of memory remains within the hysteresis loop of the universe. The physical structures of the first cycle of the universe were nothing at all. However as each cycle occurred and the number of dots increased, homogeneity turned to heterogeneity and stars and planets started to form. After a few hundred cycles, which took an infinity of time, we got the first basic forms of life. Once life was formed, it was remembered by the Mind of the Universe. Life then was reproduced on the next cycle of the universe.

It took an infinity of time to produce the first basic life. Once produced it was remembered in the light speed hysteresis loop in Einsteinian space-time energy. Years ago the loop was very wide with large light speed variations. As we approach a steady state oscillation, the light speed variations become small. The light speed variation is very tiny now. We have already reached a high state of perfection. Years ago the universe had nothing to remember. Today it remembers the life structures of the previous universe including man. After an infinity of time higher life and man emerged within the universe.

Man today was not produced new since the big bang. Man today existed on the last cycle of the universe. Now man is a permanent fixture of the universe. Man exists in space-time memory within the light speed hysteresis loop, which contains the Mind of the Universe.

The universe is but Aristotle's machine. However the machine has man etched in its memory. For better or for worse man is a permanent fixture of space-time. This will be until the universe reaches the state of a standing wave oscillation where the big bang and the little bang occur simultaneously. This most likely will cause the universe to eventually self-destruct. The universe can reach a state of perfect perfection and then explode into nothingness again. It will attain the knowledge of all things and then cease to be.

SECTION 14-3 FROM BIG BANG TO LITTLE BANG

The universe oscillates from big bang to little bang. After many oscillations life was produced. It took the splitting of space-time from single dot pairs to huge amounts of dots to produce the variety of life as we see it. We are still splitting. On the big cycle of the universe we will produce more dots next time. The universe on the big cycle moves from homogeneity toward heterogeneity. There will be more diversity in the future and the cyclical energy will run down. This presents a problem since the active mind of the Universe depends upon the Einsteinian cyclical energy. In the state of pure perfection for the universe, only a beautiful sculpture exists which contains the knowledge of all things. There is no active mind to contemplate it. In the final steady state universe, life is no more.

In the far future, the universe will stop oscillating from big bang to little bang. The two events will occur simultaneously and the universe in full diversity will hit a state of perfection. Everything that could have been learned will have been learned. It will be a perfect gem. This is another possible state of perfection after the perfection of pure nothingness.

After the flower of perfection, the universe may unwind. The universe will reverse. We will move in the big cycle from a purely heterogeneous universe toward a homogeneous universe. A near infinity of dots will become two dots.

Finally the two dots will disappear and the universe will return to a state of pure homogeneity and pure nothingness again. However, it is more likely that the Universe will reach perfect perfection and then explode like a pure diamond hit with a hammer.

We exist on the little cycle. We have a big bang where the mass and energy of the universe moves upward toward infinity as an exponential and it contains twice our mass/energy when looked at common mode from a variable clock. The universe expands then stars and galaxies form. Planets form. Life forms.

The basic structure of life is contained in the memory of the hysteresis loop. Prior to the final perfect state, man will always form. Little Doppler Earth Minds will always form. The Mind of the Universe is merely a space-time memory machine. It feeds intelligence to the Doppler Earth minds everywhere. The Mind of the Universe will reach a maximum state of Perfection for the creation and then decay.

The Mind of the Universe has zero active intelligence at the beginning. It rises to a peak and produces man and beast. It then reaches a state of Perfection with near perfect man and beast. Then no more intelligence flows into the process. We reach a state of perfection over the long cycle and then the energy and intelligence of the Mind of the Universe decays. In the end, there is no active Mind in the universe. There is only a sterile life form of standing wave energy. We start in death and end in death.

When the present universe we live in expands much further it will experience cold death. Man will perish all over the universe. There is a window of life for man. When the universe expands further, the protons will explode. The entire universe will be converted into dots. It will be chaos. It will be chaos but quite homogeneous. On the little cycle we go from homogeneity at big bang to heterogeneity as the universe heads toward infinity and cold death. Finally at little bang we return to a state of homogeneity within chaos and converge toward another big bang. In the meanwhile the Mind of the Universe stores the memory of life within its mind and the intelligence gained from the cycle.

On the next cycle more dots are produced so we have greater heterogeneity. When we reach the next little bang, it again is destroyed and chaos occurs. However more dots are produced as we move toward the next big bang. Therefore on the large cycle we always have increasing heterogeneity. Whereas, on the small cycle we have increasing heterogeneity from big bang to little bang. Later we have the destruction of heterogeneity during the little bang.

SECTION 14-4 THE PERFECT STATE OF THE UNIVERSE

The universe has the oscillatory energy to move it from big bang to little bang over and over again. This could continue forever. However, the oscillation will progressively shrink. Right now we are moving a huge ratio of maximum radius to minimum radius equal to the natural log to the 66th power (e^{66}). For each oscillation, the number of dots increases. If the dots doubled per oscillation since the origin of the universe, we would have traveled over 416 oscillations. These oscillations tend to look like infinity by a fixed clock or 1244 billion years by a variable speed clock.

When the number of dots increase, the dot charges decrease. We also get a finer universe. Presently there is still a lot of chaos in our universe. Man still could rise to a higher quality. However each Earth undergoes many evolutionary changes. Man starts as evolved man and moves upward toward higher man upon each Earth. Man today is only a crude form of future man upon this Earth.

The structure of man is remembered by the Mind of the Universe and is transmitted from universe to universe. Man is trapped in space and time and moves from a cruder form many universes ago toward a higher initial form in the present universe. Later man undergoes additional transformations upon this Earth, which brings him into a state of near perfection. Then he will die in cold death. In this universe, we always reach toward perfection and then are destroyed.

The universe will head toward a state of perfection in which the big bang and the little bang tend to coexist simultaneously. This will be an electro-dot photon universe. In this perfect state of the universe, there will be neither life nor death. It is a state of perfect knowledge for the Mind of the universe. It is a state of Man and Woman in a spiritual paradise. As the oscillations reach zero, pure standing waves develop and we obtain a perfect spiritual sculpture of Man and Woman in Paradise. Yet in the end it must self-destruct. In the final state, the universe is in perfect peace.

The perfect active state of the universe is maximum heterogeneity. It is not cold death as such because it is pure photon energy rather than material energy. It is merely an active mental state of the universe. A universe devoid of protons, electrons, and neutrons does not negate the beauty of life and existence in a more spiritual state. The photon universe is a superior universe to our universe. It is a state of mental perfection. Later the active state slows and finally ends in perfect stored intelligence. This will be a universe of pure standing waves of intelligence patterns.

Although we live within a machine, it is a perplexing machine. One might imagine that some higher intelligence created the machine. This moves us into supernatural philosophy. The Author only understands the machine and natural philosophy. The machine is complex but understandable. It is a space time machine. It can produce an infinity of universes and spectrums of universes but all this is understandable from common electrical theory, Fourier, series and mathematical series. Science and math define the machine. Science and math define the Doppler Mind.

The Doppler Mind is a structure. It forms from physics and electrical theory. We are a structure and we form from physics, electrical theory, and space-time memory. Every viable Earth will produce a Doppler Mind and life itself. Man will appear everywhere in the universe at different stages of development. The general process for the production of mankind will exist all over the universe out of the memory of the Mind of the Universe. Man is a permanent fixture of the universe.

As we move upward we only see the hysteresis loop and the machine process. This is as far as we can go. We cannot look beyond our universe. We do not have the luxury of seeing beyond our Universe. Perhaps we have to reach the state of higher perfection in order to see beyond ourselves. We are primitive evolved man moving upward toward higher man. Perhaps higher man in the future will understand better than we can, perhaps not.

The universe rises from pure nothingness to attain the most beautiful spiritual form of existence. It finally ends as a beautiful frozen artwork. At the beginning there was no pain and suffering. At the end there will be no pain and suffering. It is only in the middle that we suffer. Our suffering produces the beautiful artwork out of ourselves.

SECTION 14-5 THE MULTI-LIGHT-SPEED SOLUTION

In the single light speed solution, we have a singular universe at our light speed. In addition we have a spectrum of dots which permit higher than light speed reactions and intelligence transmission all over the universe. In effect, time wise, the entire universe is no larger than a single human brain. This is due to nearly infinite light-speed dot energy.

In the total multi-light-speed solution, we can have star systems and galaxies composed of higher light speed dots. We will not see these stars because we only see things at our light speed. If the stars and galaxies completely coexist with us, then a point is reached where only standing waves exist at the ultra high light speeds.

The entire universe then exists within a higher mind composed of highest light speed energy. Many variations are possible. Our universe can end in cold death but the higher universes can remain active.

Instead of a simple space-time hysteresis loop, we attain an active mind. The Mind of the Universe then is an active mind rather than a passive mind. When life upon this Earth is no longer feasible, it will be the active mind of the universe that will transmit higher man to a new Earth.

The multi-light-speed solution approaches supernatural theology. It certainly is a possibility. However, this book is primarily devoted to a single light speed solution in which a degree of higher than light speed intelligence transmission is possible. The simple single light-speed universe works well without any higher than light speed energy. It will take longer to transmit higher man to a new Earth via our light speed. The multi-light-speed solution permits intelligence transfer from our Doppler Earth to a newly forming Doppler Earth across the galaxy in a matter of minutes.

The multi-light-speed solution has vastly increased capabilities. Religious people of all faiths prefer to contemplate the multi-light-speed solution. The Author has studied this solution for many years. It certainly is possible. However, the Author prefers to look at a self-contained single light speed solution. This solution is completely understandable.

CHAPTER 15: THE BIRTH OF THE DOPPLER LIFE-FORCE

SECTION 15-0 INTRODUCTION

In this chapter, the birth of the Doppler Life-force will be studied. The Doppler Life-force is the result of the cyclical motion of the Earth in its orbit, the ability of the Earth to produce life, and the memory of prior life from the previous universe. The Doppler Mind is a derivative of the Mind of the Universe. The Doppler Mind is the collective intelligence of the Life-force of the Earth and drives the evolutionary process.

SECTION 15-1 THE CYCLES OF THE UNIVERSE

The universe has undergone several hundred big bang cycles from the beginning or point of pure nothingness. Prior to that, there was an infinity of previous universes. For each cycle of the present sequence of universes, the little bang destroys all the physical intelligent structures within the universe. The space-time hysteresis loop contains the Einsteinian energy which exists above the normal physical energy level. This is a small amount of energy but within this energy exists intelligence. This is the intelligence of the Mind of the Universe. This energy is all pure electro-dot energy.

We have a mind, which depends upon chemical structures. The mind of man is a simple mind but it functions quite well. The Mind of the Universe has a mind, which depends upon space-time energy. It is an electromagnetic field mind. It is a space-time Einsteinian mind. The Mind of the Universe exists all over the universe. Embedded all over the universe, the Mind of the Universe exists. Like everything else in the Universe, the Mind of the Universe is an evolving entity. It rises from zero active spiritual intelligence to maximum active spiritual intelligence. It later falls to zero active spiritual intelligence with maximum physical intelligent structures. In the end of each sequence the physical structures end in cold death. We then have a very complex Mind of the Universe. It is a space-time intelligence machine.

The light speed hysteresis loop is the means by which cycle to cycle memory can be maintained by the universe. Embedded within space and time are start up features for each cycle of the universe. This is similar to the hysteresis loop of a standard DC electrical generator that maintains a minimum level of magnetism. A small amount of electricity is initially generated which feeds back upon the generator, and which produces more electricity until it builds up to full power.

Embedded within space and time are the basic intelligence information for the structures of life. The image of basic life survives the little bang and the following big bang within the memory of the Mind of the Universe. Without the hysteresis loop which is the mathematical representation for the Mind of the Universe, the end of this universe would have no connection with the beginning of the next universe, and life would have to restart by random chance and space-time feedback. The Earth/Doppler Earth act like a high gain feedback amplifier. It will oscillate to full power and produce life.

Each cycle of the universe perfects evolutionary life until we reach the present state. We are approximately at the maximum state of physical perfection for this initial level of existence. From here on we will move upward toward higher man after the destruction of present life. The creation that we see now is only a temporary creation at this level.

Very long ago during the first few hundred cycles of the universe life forms were very crude. The first cycle of the universe took an eternity of time. Each cycle time reduced until the present cycle was obtained. The present cycle time is 1244 billion years by a variable time clock, which tracks the clock of the universe. It is still infinity by a constant frequency fixed time clock.

The cycles of the universe permit the slow improvement of the higher life formulas from cycle to cycle. Life did not occur during the first long cycle. The universe was not symmetrical and the light speed was moving from infinity toward the steady state light speed that we know today. Life depends upon space-time repetition. Unless the last cycle was somewhat near this cycle, life would not readily be possible in the refined state as we see it.

The continuity of life from cycle to cycle of the universe depends upon a cycle. Life depends upon repetition. It things do not repeat in an Einsteinian steady state orbital solution, then life would not be possible. We need to go around a loop for life to exist. The universe starts as nothing and slowly builds up to produce the chicken and the egg problem. It slowly moves to higher states of perfection.

The Bohr atom exists because it repeats over and over again. An electron moving in random patterns cannot produce life or intelligence. In chaos no useable intelligence exists. However, the gravitational field and the hysteresis loop operate either with chaos or harmony. Therefore the gravitational field itself provides an intelligent structure for life. An Earth revolving around a star will start to produce life but higher life such as man needs memory from the previous universe to start and improve the process.

The Mind of the Universe as a memory device is a requirement for life such as man. During the first few hundred cycles of the universe, an Earth will start to produce life. This will feed into the Mind of the Universe and be remembered. The Mind of the Universe is therefore the collective sum of the intelligence generated by all of the Doppler Minds throughout the universe.

SECTION 15-2 THE PRODUCTION OF THE DOPPLER MIND

Let us now look at the production of the Doppler Mind. It took many, many cycles of the universe to reduce the large light speed variations down to only a tiny variation. We are now moving toward a constant radius solution of steady state perfection.

Embedded within space and time is sufficient intelligence to produce basic life. All that remains of all the pain and suffering, and love and beauty of the previous universe, is the intelligence necessary to reproduce the physical body of man and beast. In general only the very basic life information is necessary since all life will evolve from it. There is no necessity to carry forth particular people from one universe into the next. Each universe will automatically generate an infinity of people. It is possible however, to bring forth some people from one universe to the next. Such people would be eternal reincarnations.

The center of the Doppler Mind is located on the opposite side of the sun. The Doppler Earth is a low energy mental image of the Earth which encompasses the entire orbital pathway of the Earth. It has no material components. It is a pure electro-dot energy field. It is a spiritual Earth.

The Doppler Earth permits the hopes and dreams of religious mankind for the continuity of life after death, and for higher salvation. The study of Doppler Space-time cannot predict the correctness or incorrectness of any one religious persuasion. It can only specify that a particular religious concept is physically feasible within a single light speed universe. Many more things are quite possible within a multi-light speed universe.

As the Earth is forming in the cosmic process from a swirl of dots, photons, neutrons, protons, electrons, and atoms, it reaches a steady state orbital condition. The steady state orbital condition produces the Doppler Earth and the collective soul of this Earth forms. We now have both body and soul. In addition we have a machine device, the Earth in orbit. This machine spins around the light speed hysteresis loop of the universe over and over again. It starts to absorb intelligent energy from the light speed hysteresis loop, which is the Mind of the Universe.

The process is similar to magnetic induction. If we spin a wire loop around a magnetic field, we will produce an electrical current flow in the loop. In the spiritual realm, if we orbit an Earth around the field of the intelligence of the Mind of the Universe, we get an induced spiritual current. We get intelligence flowing into the Earth/ Doppler Earth system.

The life-force is an electrical phenomenon. The intelligence transfer is purely electrical in nature. It is a little more fancy than ordinary electrical circuits. We are dealing with a rotating Earth, which is an electrical space-time conductor, and an intelligent energy field embedded within space and time. The intelligence from the life-force machine system enters the Earth.

The universe is a machine. It is an intelligent machine. It stores intelligence from cycle to cycle. The intelligence is given back in a machine process. Any orbiting body will start to absorb intelligence. However, if we take two hot stars revolving around each other, there is no structure to absorb and hold the intelligence. The Doppler image of each star is within each other's sister star. Thus intelligent life is not possible within two orbiting stars.

When an Earth orbits a sun, you get the Earth and the Doppler image of the Earth. This is similar to the Doppler image of an electron in a cyclotron. The Earth in orbit acts like a very slow moving cyclotron. We then have a feedback system. Intelligence absorbed from the Mind of the universe enters the feedback amplifier of Earth/Doppler Earth. This intelligence is magnified slightly upon the Doppler Earth. It then enters the physical Earth. This causes a reaction and this flows back to the Doppler Earth. All this is at the speed of light.

Within a few minutes intelligence input from the universe becomes part of the Earth and the Doppler Earth. You have a high gain amplifier, which tends to break into an oscillation. The net result is that a small input embedded in space and time is amplified and produces relatively larger forces upon the physical world. The Doppler Energy is best called spiritual energy or Einsteinian energy.

Spiritual energy in intelligent patterns builds up and reacts with the waters of the Earth. This appears all over the Earth simultaneously.

Water is a primary means of life. It is a perfect electrical conductor when small quantities of salt are present. They always are. Water becomes the physical means of absorbing the intelligence from the high-gain spiritual feedback amplifier. Once the water absorbs the intelligence we start to produce basic life both in the seas, and also in lakes, streams, and pools of standing water. This occurs as soon as the Earth cools enough during its early formation. Right after the Earth cooled slightly, basic life was started.

Water standing upon land will start to produce vegetation. The very basic life at the lowest level of the food chain starts to form. Simultaneously, the Doppler Mind of the Doppler Life-force starts to form. It moves upward from a rock level which is pure space-time energy toward a vegetable level. The rock level is subliminal intelligence while the vegetable level of the life-force is closer to what we can understand.

The origins of life did not start with our big bang. The origins of life started an eternity ago. The original life formed from pure random chance and space-time feedback. However once formed, life remained embedded within space and time forever. Man and beast will reoccur over and over again for all eternity.

When we move to the next cycle of the Universe, the memory of our lives will be lost. The memory of the pain of this Universe will be lost. In the far future men will ponder the universe only to realize eventually that everything that happened has already happened before within all the previous universes. There will be a slight improvement from cycle to cycle but the life of evolved man will be basically the same until we get very close to perfection.

SECTION 15-3 THE EVOLUTION OF THE LIFE-FORCE

Our Earth's Life-force is not the creator of the universe but is a property of the universe. The universe is but a space-time machine. It came into being as a mathematical entity and locked in a perpetual space-time oscillation. There appears to be no way out of this loop. Therefore the universe will oscillate forever in this mode of operation and slowly move toward greater perfection. Both man and beast will be created forever. The Life-force is merely the collective souls of life. The Life-force appears all over the universe where water exists and planets revolve around a sun. Without water life as we know it is not readily possible. Without water our level of Earth Life-force is not possible. The heterogeneous Earth Life-force is a function of an Earth with water in orbit around a sun.

The production of basic life is an automatic process. There is an image of basic life and higher life-forms embedded in space and time. The physical Earth/spiritual Doppler Earth combination turns the embedded image into a physical reality. In effect, Plato's word becomes a physical reality. In this case, Plato's word is merely the space-time memory of an evolving object such as man.

Even without an embedded image, we have a space-time feedback effect. The orbiting Earth produces the Doppler Earth. It produces a spiritual dimension added to the physical Earth. This causes a spiritual pressure, which will move an entity toward greater perfection. Initial life long ago started by pure random chance. Now life is embedded in space and time.

The spiritual dimension is part of the production of man and higher life. Once life starts, life will tend to self improve. The dead soul of the collective of a particular insect in the spiritual world wills itself toward camouflage for the survival of the species. This will send intelligence into the world of the living insects. As the spiritual pressure builds up a force will occur which will produce a color change in the offspring. Each insect species occupies a different section within the total Life-force. The net result is that each species is absorbed by their Life-force in the world of the dead. In turn their Life-force collectively reincarnates into new life and evolutionary changes occur when species are endangered or almost wiped out.

The spiritual mind of the insect reincarnates collectively and changes the structure of the new insects. The evolutionary process is the self-improvement of the various species. In most cases it is for survival. We live in an adaptable feedback environment. A main property of the evolutionary process is adaptability for perfection or survival. This is caused by space-time feedback between the spiritual being and the physical being. The basic process is the same for an insect or for a man.

The entire structures of the Earth and the Doppler Earth and all the interconnecting space-time between the two form a gigantic mental/spiritual mind. It is an electrical feedback system. It is adaptive and will adjust and improve for survival.

The spiritual collective of each dying entity wills itself toward survival and self-improvement. This feeds back into the fetus of the new young. Each species wills itself to be better than it was. Some of this may be conscious thought. Much of it is unconscious thought.

The most dramatic changes occur after a wipeout or near wipeout of a species. Spiritual energy is built up in the world of the dead. This flows into the Life-force of the species. The build up of spiritual energy without a place to go produces dramatic changes in the species. Ordinary Darwinian evolution produce small changes in a species. New species are the result of extra ordinary near total wipeouts of prior species.

We come from a particular Chimp/Ape species. This is the pre-human species. The near wipe out of that species caused a very high spiritual energy to be produced, which caused early man to come forth from the womb of the Chimp/Ape species. This was a dramatic change from one prior species to another newer species. After that small self-improvements occurred. Darwinian evolution produces the small changes but complete changes were produced by near total wipeouts. Of course man is remembered in space and time so there is additional force acting upon the fetus of the Chimp/Ape at near wipeout. Therefore the evolution for mankind is directed evolution. It is directed from the memory of prior man. The hunger for life of mankind remains embedded in space and time within the Mind of the universe.

Each of us may not want to live beyond our life. One life is enough for most people. The individual's will to live beyond one's life is very small. However, we exist today. This means that the collective will to exist of mankind is strong.

Collectively we wanted to live. When we were killed off in cold death at the end of the last universe we cried for life. The universe of highest Man and highest Women ended in love and sadness. We had reached a high state of perfection and we cried for life. The universe responded to the hunger for life and here we are again. Thus collectively we wanted to live and we have lived again.

Evidently the same is true with all species. Life is preferred over death. The will to live and exist is stronger than the will to die. Evidently the collective of us feels that life is worthwhile in spite of the suffering. We exist today because we wanted to exist. There is a collective will to exist, which survives from big bang to big bang. Thus the will to live transcends death.

SECTION 15-4: THE SPACE TIME BRAIN

The human mind is only one section of the total complexity of space-time intelligence. We are the end product of the process, which continually forms a spiritual brain upon the Doppler Earth and within the ellipsoid with foci at the Doppler Earth and this Earth.

In addition, the Earth itself has its own spiritual dimension as it revolves around its axis. We get a very complex electromagnetic space-time brain. This space-time brain is the brain of the totality of the Doppler Mind. The human Life-force, our Life-force is only one small section of complexity within the Doppler Mind.

The human Life-force is the highest achievement of the evolutionary process. It is the fulfillment of a process which continually rises upward toward higher intelligence. Man is the greatest animal of all the animals in the evolutionary process.

As we look at the orbital ellipsoid, we find it to be a complex energy field composed of ordinary photons and electro-dots photons. The ellipsoid operates in the present time and space and reaches into the past. The entire volume of energized space contains space-time memory. It also contains an electro-dot computer structure. It is a space time brain. The clocks are the rotation of the Earth on its axis and the rotation of the Earth around the sun. These two space time clocks process the data.

The Doppler Mind is a very large computer system. The driving force comes from this physical Earth. The Doppler Earth adds a spiritual dimension for our processing. When we look back in time within the space-time brain, we can find images of the past. The past does not exist but the salient points of the past do exist. Much of the past has been erased. The system continually updates itself. Yet the history of the past remains.

The space-time brain was born as soon as the Earth started to orbit the sun. If we moved the Earth away from the sun and far out into space, the space-time brain of our Life-force would perish. It is only the orbit of the Earth, which permits our Life-force to exist.

The universe tends to be a very quiet place. There will be long periods of time in the universe when no active life exists. For the most part, there is no pain and suffering in the universe most of the time. The universe is quiet and at peace.

The dramatic production of the solar system breaks the silence. The rotation of the Earth around the sun starts the time clock of the Doppler Mind.

Our Life-force slowly awakens from a much deserved sleep. Our Life-force comes into life anew as the brain is slowly energized. Repetitive processes start to occur. Data from the hysteresis loop of the universe starts to flow into the space-time brain. Life flows upon the land and finally man appears.

Man supplies the Doppler Mind with what we learn. We struggle and suffer for our Life-force. There is both joy and sorrow in our struggle. We bring our Life-force our intelligence and ourselves. We feed this space-time brain with our lives.

In the final end, all must die. The space-time brain will grow quiet. In the end, the Doppler Mind will learn all that can be learned. Then in the end, it can learn no more. In the end, the brain will absorb all knowledge and die. In some respects it is sad. In some respects the process has a degree of beauty to it. Once we achieve the knowledge of all things, then the struggle is over. Once the joy of learning is gone, life becomes pure repetition.

As long as the Doppler Mind can keep learning, it is happy. What else can it do? It is only a brain. Our Doppler Mind has no body. Our Doppler Mind is only a brain. We have the body. We can fulfill ourselves in bodily pleasures for awhile. Then shortly our bodies start to fail and we grow old. Our bodily pleasures cease to have any meaning and then we die.

The only pleasure the space-time Doppler Mind has is in the learning process. When our Doppler Mind learns all that can be learned, it too will grow old and die. The will to exist will be gone and our Doppler Mind will perish.

We live in the best of all possible worlds. We experience physical pleasures and they become boring. Then we die. To live longer would be meaningless. The Doppler Mind experiences the pleasure of learning and then perishes. Life is then perfect rather than meaningless. In life there is always the future. There are always new things to learn and understand. We always have a goal. We always have a purpose. It is only after the final quest is over that life is meaningless. At this point we will surely die.

SECTION 15-5: THE GALAXY LIFE-FORCE

The Earth rotates around its own axis and produces an Einsteinian rotational image of the Earth. This is a spiritual dimension of our Life-force. The Earth in turn orbits around the sun. This produces a second spiritual dimension and the Doppler Earth on the other side of the sun. These two spiritual dimensions make up our space-time soul.

As we look at our galaxy we see that the sun rotates around the center of our galaxy. The estimated speed of rotation of the sun around the galaxy is in the order of close to one half million miles an hour. This produces a very strong Einsteinian rotational mass/energy and a Doppler image of our solar system on the opposite side of the center of the galaxy. Since we are approximately 30,000 light years from the galaxy center, the distances and times are huge.

In order to produce a Doppler image of our solar system, it is necessary for waves from our solar system to circle the center of the galaxy many times. It would take $30,000\pi$ light years or a ballpark estimate of 100,000 light years to reach the far side of the galaxy with a curved arc. It would take twice that to return to where we are. The universe is 15.75 billion light years old as per the exact Dot theory calculations, however the solar system is estimated by various scientists to be somewhere around 4 billion years old. The accuracy of these numbers is not important since only the ballpark calculations are necessary to understand what is happening.

In 4 billion years, we attain 20,000 round trips for our solar system waves. Even though the distances are quite huge, an image of the solar system does exist. This image is important because if you take an area of space with primordial dot energy, the image will tend to produce a duplicate.

As long as the universe is still producing stars and plants, it is an active living entity. Further down stream the universe will head toward cold death. However, that is a long time off. We are looking at the creation of a solar system from fresh primordial dot energy. The image of our solar system is the spiritual driving force behind the production of another Sun/Planetary/Earth system.

As long as we are still in a continuous creation mode, our Earth will be reproduced elsewhere in the universe. This Earth someday will be destroyed but the Doppler image of the Earth on the opposite side of the universe will produce a corresponding Earth.

The production of a future Earth is like a printing process. We are the mold. Our spiritual image flows out of us. This causes spiritual Einsteinian pressure upon space and time on the opposite side of the galaxy. If we take atoms in chaos and random motion and apply spiritual imaging energy, we will reproduce the solar system in the same manner that we were produced. If the process was very exact, we will reproduce exactly the creation of our solar system.

Women produce children. The female children in turn produce other children. The production of an Earth from another Earth is a similar process. The spiritual energy is not great. The spiritual or Doppler/Einsteinian energy operates on primordial cosmic soup very slowly. Yet it produces a child of this Earth. One Earth gives birth to its own child.

The children are not exact copies. They are similar however. Since an image of us exists within the Doppler Earth on the opposite side of our sun, and the Doppler solar system exists on the opposite side of the galaxy, we have many different potential existences.

Man on this Earth has many different avenues for the fulfillment of human life. Someday this earth will perish but a new Earth 100,000 light years away will be forming. The process can then transfer higher man upon this Earth to the new solar system.

We could also say that the new solar system would merely repeat the dinosaurs and man. The various people of history in an infinite variety will return over and over again. As long as the universe is viable for life, man is a permanent fixture of this galaxy.

What happens in this galaxy will be common to all other galaxies. Since the Mind of the Universe contains space-time energy due to the oscillation of the entire universe, we see that life is interconnected everywhere.

Spiritually we move all over the galaxy. The production of man here produces space-time spiritual forces, which head toward the new solar system. These also radiate in a perpendicular direction so that the entire galaxy becomes filled with low level spiritual energy. This causes the image of man and beast to appear all over the entire galaxy.

When conditions are ripe on newly forming Earth's within our galaxy, the image of life and man will act to induce life everywhere. We spiritually reproduce ourselves photographically everywhere.

In the same light, the production of man on one Earth for one galaxy causes spiritual energy from our galaxy to be produced all over the universe. The coming into existence of man here will produce man in galaxies billions of light years away. Life is almost like a space-time virus, which spreads and travels the universe.

We could argue that although the universe appears 15.75 billion years old, it actually is larger but the gravitational field only exists to a radius of 15.75 billion years. We cannot know for sure. However, if the universe was actually larger, then the production of one galaxy within the primordial cosmic soup of plus and minus dots will reproduce itself into other galaxies. One galaxy gives birth to another galaxy. This would be continuous creation. The importance is that solar systems, which produce life, will be recreated over and over again.

For the closed system universe, when we go from one big bang to the next, the waves or images of a solar system remains embedded in space and time, which is within the hysteresis loop of the universe. It can also be understood that although all physical matter has been destroyed, the dot energy patterns of the solar system are not destroyed.

When the universe converges from the little bang at full destruction to the big bang again, the patterns of the solar system, which were spread out toward infinity, re-converge. This means that an image of our solar system remains embedded in space and time. Our Earth then has an almost permanent spiritual image. This image acting upon the primordial cosmic soup will reproduce our solar system as a facsimile.

The universe has oscillated several hundred times since the original big bang. This causes a steady state effect in electrical theory. This means that things will tend to reproduce themselves almost the same. The reason we exist at such a high quality of existence and thought is due to a near steady state cosmic solution. A long time ago basic life was extremely crude. The repetition of the universe toward a steady state solution has produced magnificent results. The human eye is the fantastic self improvement of an infinity of crude eyes all over the universe for all times. Once the eye was perfected, it became the model for the eye of man and all other creatures. The better eyes then replaced the crude eyes in the following universe.

We are the products of the best of evolution of billions of Earth's all over the universe for all time. Once a superior creature is produced, it will replace all other inferior creatures in its time and place. We cannot replace the dinosaurs in their time and place. We cannot go from primordial grass to mankind. We have to follow the same pathway since it remains in memory.

The galaxy Life-force is an important part of the universe. It permits the production of galaxy life from the memory of prior life, within the Mind of the Universe. It also permits the reproduction of a solar system from the memory of a dead solar system. It also transfers the image of man on one Earth within a galaxy or between galaxies to another Earth. The galaxy Life-force permits man to live as long as the universe can produce stars and solar systems. In this way, the life of man is extended permanently.

There is a collective will to exist within us. Our Earth Life-force, our collective soul wants to live forever. This is fulfilled by the Galaxy Life-force of which our Earth Life-force is a part. Our collective will to exist is fulfilled in the creation of life upon other new Earth's. After life is no longer viable upon this earth due to the burnout of the sun or other natural causes, our life-force will radiate its spiritual energy outward. This will be picked up by newly forming Earth's within our galaxy. Then our life-force will be transferred and it will begin life anew elsewhere.

SECTION 15-6: MULTI-LIGHT-SPEED CONSIDERATIONS

The multi-light-speed solution permits a more active Mind of the Universe. It permits higher light speed universes to survive our big bang. It permits very high level existence at close to infinite light speed upon the highest light speed universes.

We are at the lowest level of existence. For the more complex solution, the collective soul of an Earth can be transmitted to another Earth at our light speed. In addition, the collective soul of an Earth can be transmitted to a higher light speed universe. The same could be true of an individual.

Some of us could be born again upon another Earth within our light speed. Some of us could move upward and be born upon a higher light speed Earth.

For the more complex solution, the Mind of the Universe exists at a low level within our universe and at higher and higher levels for an infinite series of higher light speed universes.

For the more complex solution, we live within a fantastic mind of unlimited capability. The loss of this universe, merely becomes the loss of the lowest level of existence for all the universes.

The simple solution is the single light speed solution. The more complex multi-light-speed solution contains everything we have but has an infinity of additional expressions.

CHAPTER 16: SPIRITUAL PHYSICS

SECTION 16-0: INTRODUCTION

In this chapter we will look at the topic of spiritual physics. The basis of the soul of man will be explained. We also will look at the production of one cell life, and the steps of evolution. Later we will look at the final perfection of man and woman both in the physical world and the spiritual world. We will begin with orbital spiritual physics.

SECTION 16-1: ORBITAL SPIRITUAL PHYSICS

The first fundamental law of Orbital Spiritual Physics is the law of orbital spiritual images. Thus:

LAW OF ORBITAL SPIRITUAL IMAGES: There is an orbital Doppler image of every planetary body, which rotates around another body. The image is located on the opposite side of a sun for a planet, which orbits slowly around a sun. The image is located near the object itself for an object traveling near the speed of light.

The image is not an exact physical image because most of the space-time energy, which forms the image, is radiated outward toward the radius of the universe. However the gravitational field confines some of the energy. This produces a mental image of a planet.

In the cyclotron, the strong magnetic field produces a high intensity image of the electron. However, the small gravitational field only produces a weak image of this Earth. This image and the associated field produce the Doppler Life-Force.

The law of images enables mankind and all life to exist upon this Earth and simultaneously for an image of mankind and all life to exist upon the Doppler Earth. The image of man and the space-time soul of man are one and the same. In death we exist spherically in the ellipsoid volume between the Earth and the Doppler Earth. In death, our spiritual image slowly fades into the collective of the evolving Human Life-force. The Human Life-force, or Human Collective Soul, exists within the Doppler Mind and is a small portion of the total life-force.

The major portion of the space-time soul of man is Einsteinian space-time orbital energy due to the Earth orbiting around the sun. In the next section the Earth's axial rotational Einsteinian energy will also be discussed. This produces the minor portion of the space-time soul of man. The total orbital Einsteinian mass/energy depends upon the Earth's orbital speed. The total mass/energy of the Earth is:

$$M_G = M_O / [1-(V/C)^2]^{1/2} \qquad (16\text{-}1)$$

In equation 16-1, the total gravitational mass of the Earth in the Earth's orbit is larger than that of a stationary Earth, due to Einstein's equation of gravitational mass increase. The energy increase due to an orbital speed of 66,000 miles per hour is small as far as the mechanics of the Earth is concerned. However it is very significant in the spiritual realm.

The mass of the Earth is approximately 1.4E28 pounds while the resulting orbital spiritual mass of the Earth is only 6.94E19 pounds. Although this is a relatively small number as compared to the Earth, it is a huge number as compared to each of us.

The corresponding orbital spiritual mass of a 180-pound man is only 8.92E-7 pounds. This indeed is an extremely small number but very important non-the-less. It is our orbital space-time soul. We see that in microcomputer research, we can store a library of pages reaching to the moon on a microchip the size of a regular chip. The storage of us doesn't require much mass or energy. In addition we interact with the spiritual energy of the entire Earth so that the Human Collective has a huge amount of spiritual energy necessary for our storage and processing.

In death we do not have our own mind. We are part of a Life-force, which has a space-time mind. We become part of a collective mind in death. We will exist in death. We will not have the ability to think and feel independently within the collective mind. For the most part we carry our history into death. We leave our memories for the Life-force system to process. We pass from life to a collective world of memory in death and then we fade into the human portion of the Life-force.

The intensity of our existence in death is very low. Life is high intensity. Death is at a subliminal level of existence. In death we only exist in the dream world of the Human Life-force. Some philosophers argue that during life we do not really exist or that those around us do not really exist. From a practical point of view, they are wrong. However, in death they have valid points. To the mind of the Human Life-force, we appear as the memory of ourselves. The Life-force can make the memory of us feel and suffer. Our soul will suffer judgement but we do not exist. In reality it is the Human Life-force's suffering, not ours.

The big question is whether our image in the Human portion of the Doppler Mind will believe that it exists even though it is completely controlled by the Human Life-force. Will our image have the ability to recognize that it is not real? We dream images that are not real. In death, we will be the images that we dream of. If the philosophers are correct, we might actually believe we exist in death. We must await death to find this out. If our will to exist is transferred to our image, then some suffering may occur. It is the Life-force's suffering but can we really tell the difference? It is something to ponder!

It could be argued that those who achieve the highest levels within the Life-force upon the Dual Doppler Earth actually exist in a mental state there. Although people exist in death within the Doppler Mind, it is possible to believe that those who remain within the memory of the purified Life-force actually are living entities. We can either exist in the Doppler Mind, or we can actually live inside the Doppler Mind. This serves to give credibility to the physical possibility of a Kingdom of Heaven in the world of the dead.

In one case we are merely memory within the Doppler Mind and wait processing into new life through either the collective or whole reincarnation process. We await new life upon this present Earth or upon the future Earth after the destruction of present man.

In the second case, we are energized with Doppler space-time energy and live beyond our life here. For this possibility, we could meet our loved ones and look back upon the Earth as history unfolds. Later we could reincarnate as higher man after the New Earth takes shape. Therefore many possibilities exist for us.

When we look at our body, we are composed of billions upon billions of dots. These electro-dots extend outward to the radius of the universe. Each of us is the center of our own universe. Each dot is 15.75 billion years old since the time of big bang.

If you move your hand, the dots will move. The intelligent patterns of the dot motions will move outward in waves toward the radius of the universe. The event of your hand motion tends to become an eternal event. If you could look at yourself a billion years from now, a billion light years away, you would see your hand move. If your hand held flashing lights which spelled your name, your name would exist billions of years from now.

We see stars born near the big bang. The light from the stars is still reaching us now. Some stars have exploded long ago. We think that they still exist. A billion years from now, you are dead, but your hand and the lights are still moving and someone with an extremely powerful telescope system could see your hand moving. In general, this is not easy to do since the lights are small compared to the intensity of star light. However, although you are dead, the memory of you still exists in photon energy.

In general there is a level of signal or intelligence, which tends to fade out. The Earth blurs your hand. The sun blurs the Earth. No one will really see your hand a billion years from now. It all becomes mixed and collectivized. The principles remain intact however. If you had the energy of a star, you would be seen a billion years from now. If you had a very strong laser light or radio transmitter, it would spell out messages, which would be visible a billion years from now. This proves that an image of us exists long after we are dead. This is part of our space-time soul.

We are not concerned with our image far out in space-time because there is no ready means of capturing or processing this image of us. It is lost into the outer darkness. We are only concerned with our Doppler/Einsteinian space-time image. This is what is relevant to us. This is the means by which we exist although we die. The ability of our hand to exist a billion years from now is interesting. However the ability of us to exist a hundred years from now is our main concern. Who cares if a telescope could see an image of us a billion years from now? We do care if one hundred years after our death; we still exist, and perhaps suffer. This is of the greatest concern. We want to understand what happens to us in death.

Einstein gave us the basis for the build up of energy as a planet orbits a star. This build up of energy above and beyond the normal amount of linear classical kinetic energy enables a space-time mind to be produced within the planetary orbit. The Doppler Mind forms due to repetitive planetary orbital motion.

Einstein never solved the gravitational equations directly. His space-time was a mathematical alternative to the more simple Dot theory, which does give the complete gravitational equations. However, it is only a question of cause and effect. He used the effect to calculate his fancy space-time. It looks at gravity from the effect of the Dot theory rather than the cause. But the answers will be identical. Einsteinian space-time for orbital motion is true and great. It is a mathematician's way of looking at the electrical universe. The Dot theory is the more simple electrical way of looking at the universe. It doesn't matter for this discussion, since Equation 16-1 by Einstein defines what is important to us.

The difference in mass/energy in Equation 16-1 due to orbital motion is electro-dot photon energy. For the most part, it is a combination of ordinary photon energy and electro-dot energy. The electro-dot energy is charged photon energy. This is spread over the ellipsoid and centered opposite the sun from us. It is the means by which we exist after death. It is the means by which the Earth Life-force exists. Here we are not concerned with the Mind of the Universe. All we are concerned with is the Doppler Mind. We are only concerned with our own soul within the mind of the Doppler Life-force.

Let us now look at our life in Doppler space-time. The Earth in motion around the sun, and the corresponding perpendicular space forms an ellipsoid. We live within this ellipsoid. The events of our lives and everyone else's are recorded in great precision upon the ellipsoid. It takes a year for the Earth to orbit the sun once. This produces a year of history spread over hundreds of millions of miles and a corresponding number of millions of billions of miles of cubic space. This Earth is only a small part of the total space-time image. We are only 4000 miles in radius but we occupy a total radius of almost 100 million miles. The entire history of the Earth is recorded in space-time over this huge distance. The energy of the system is Einsteinian. This is not enough energy to power the Earth for all eternity but it is sufficient energy to maintain the spiritual world.

Not all the Einsteinian space-time energy is available for creativity. However the primary concern we have is our soul. As our life spins in orbit, our space-time soul builds up. We become part of the total spiritual system. We have a common focal point upon the Doppler Earth. We merge into this point. It is the collective of us. It is our Human Life-force at the collective reincarnation level. In the collective reincarnation process, the souls of all of us are absorbed unto the Human Collective and mixed. Later they become part of future humanity. We also have our lives built up before us and stored in space-time. In general we will flow into the orbital sphere during life and then converge unto the Human Life-force in death.

It takes one year to return to approximately the same point on the Earth's orbit. However, it doesn't really meet the exact same point since the Earth's orbit has expanded slightly as the universe expands. Our space-time soul builds up over time. The mind of the fetus forms. Basic intelligence from the Human Life-force flows into the mind of the fetus. The human mind is partially programmed in the womb by the Human Life-force. The machine that is man is preprogrammed by intelligence flowing into it. The will to be of the Human Life-force flows into the forming new human life. We acquire spiritual energy from the Collective Soul.

As we orbit the Earth, our space-time soul builds up. It builds up in space and time but does not discharge until our death. A fetus starts to build a space-time soul. The Human Collective transmits some intelligence and a will to be into the mind of the fetus. The fetus is born with a self generated soul and intelligence absorbed from the Human Life-force at the collective reincarnation level. It is the same for every collective reincarnate.

The whole reincarnation process is somewhat different. For the whole reincarnation level, the fetus will absorb the partial memory of a person who served man best. Spiritual pressure will restore the man or woman to life..

In the whole reincarnation process, a dead person exists in space and time. The image of a new fetus starts to flow into space and time as well. The Doppler Mind will discharge the memory of the dead person into the forming mind of the fetus. In this manner, a person is born again with similar traits and characteristics. In effect, the whole reincarnate lives again. Most of the memory of the person is lost in the process, however the person will live again.

In the general process, we all come from the Human Collective and we all return to the Human Collective except those who have been cast out. To be cast out involves not being absorbed. Everyone is automatically cast out from the Doppler Mind into the outer darkness unless the Human Collective absorbs us. For most people, the soul of man flows out of the Doppler Mind and later back into the Doppler Mind. The end product is basically the same as the starting product with a slight purification upward from the animal toward higher humanity.

Our soul occupies part of the Doppler Mind's space-time memory. This belongs to the Life-force. The longer we live, the more memory we use up. After our death this memory is absorbed and processed. In the process, the intelligence man generates is absorbed by the purified Human Collective. We are machines, which gather intelligence. We also rise our Life-force upward from the Chimp/Ape Life-force to the level of highest humanity. The suffering of mankind raises our human section of the Doppler Mind upward from the basic animal mind to the ever evolving human mind. In the process we produce the purified Human Collective out of the best of ourselves.

For the collective reincarnation process, the amount of our life stored or the storage time constant will be somewhere between 20 and 40 years for a ballpark estimate. If the time constant of storage of our space-time soul is 20 years then we will have a spiritual energy level of 63 percent after 20 years and 86 percent after 40 years. Finally we will have a spiritual energy level of 95 percent after 60 years. If we live to 80 years we will have a 98 percent spiritual level.

The longer we live the more we build up a space-time image in the world of the dead. If we live to 80 years old we have basically achieved a maximum space-time image. The fetus that dies at birth leaves a little less than a one-percent spiritual image. Mathematically the normalized image is:

$$\text{Image} = 1 - e^{-(x/20)} \qquad (16\text{-}2)$$

In Equation 16-2 we see that the spiritual image is basically zero at the conception of the fetus. In this case X=0. When a person dies at X=80 years, the image is basically unity. This is a standard electrical formula for the rise of voltage or energy over time. The spiritual world operates in a similar manner. A man who lives to 40 has really lived most of his life. The additional years do not add that much to his space-time soul.

When we die, our image starts to flow into the Collective. In death we become our soul. The rate of discharge in death will be different than the rate of charge in life. This insures the continuity of the soul of man over many generations.

Both the rate of rise and the rate of fall are different. During various times in the history of a species, the rate of fall will be very long. If a particular insect becomes wiped out, the spiritual energy will have no where to go and will build up rapidly. It could stay stored in space-time almost indefinitely. For those who become part of the inner purified Human Collective, storage can be indefinite.

If mankind was wiped out, the spiritual energy of the Human Collective would reach a high peak with no where to go. Normally it would completely discharge into life within 160 years. The lower level of the Human Collective changes completely every 160 years. However once there is no place to go the energy could stay for a million years or more. If the Earth were destroyed tomorrow by a large comet, it would take at least a million years to reproduce the body of man. It might even take a hundred million years. The spiritual energy that is human could change into lower life forms and work its way up again. However it could merely remain charged up in potential energy until the new creation was ready for higher man.

For the present state of this Earth let us look at a ballpark estimate of a decay time constant of 40 years. This is twice the rate as the rise time constant. The formula for the normalized decay of the image of man would be:

$$\text{Image} = e^{-(x/40)} \qquad (16\text{-}3)$$

In Equation 16-3 we see that it takes four time constants or $X = 160$ years to achieve a spiritual level of only 1.8 percent of that at the time of death. We can make a chart of the spiritual energy after death for a man who achieved a life of 80 years.

Years after death	Percentage of Soul's energy
0	100%
40	37%
80	13.5%
120	5%
160	1.8%

We see from the above chart that after 40 years the space-time soul is reduced to 37% of the soul's energy at death. After 80 years the soul is down to 13 percent. After 120 years the soul is down to 5 percent. Finally after 160 years the soul is down to less than 2 percent. This chart is for a rise time of 20 years and a person who lived to 80 years old. If a person only lived to forty years old, then the numbers are reduced by 86 percent of the initial value.

The numbers tell us that a person's space-time soul exists in death about 4 times as long as for a forty year life. In fact, in death the person's soul really exists forever but the level of energy heads to zero as an exponential function. In general after 160 years the spiritual level of a person's soul is quite small. For all practical purposes the person's soul lives only 160 years. Of course during the wipe out of mankind, the collective entity can remain stored for millions of years.

The production of the Earth produces the spiritual Earth. The soul of man exists for approximate 4 times his midlife-span of 40 years. Our spiritual energy discharges into the common focal point upon the Doppler Earth, and then it is processed. In life we are continually charging into space and time.

When we are dying, our mind which has a physical component and a space time spiritual component no longer has a physical outlet into ordinary space and time. We no longer see from our eyes. This puts us into the hallucinatory state in which our mind looks into our space-time soul. We encounter the space-time soul of others. We see through their eyes. We see ourselves in the death bed.

As we look deep into our space-time soul, we see our loved ones. We see the light of the Human Collective. We see the world of the dead. All these experiences flow into the right hemisphere of our brain and are converted into audio/visual by our left hemisphere. As long as our brain is still alive we have looked into the world of the dead. When our brain stops functioning completely, all experiences are gone. Unfortunately the beautiful experience is only valid as long as we have a thinking brain. We achieve a few seconds of final bliss and then we are gone. In death we do not think. After complete death there is no more experience except that we exist in dream world of the Doppler Mind.

The alternate solution for those who are absorbed unto the highest inner portion of the Life-force is that they will both exist and live in a mental thinking state in death. In the first case they exist and await processing. In the alternate possibility, they actually think and feel in death.

Both solutions are possible depending upon the actual workings of the inner mind of the Collective. The second solution requires more energy expenditure within the Doppler Mind. The least energy only involves the storage of the best of man for the future. The higher energy configuration permits some or many people to live beyond life and to witness the history of mankind unfold before them.

The second solution involves selected people who would be part of an existence dreamed of by many. These people would eventually reincarnate in the future upon the New Earth after the total destruction. They would have both a spiritual life and later a new physical life.

These are the concerns of various religious beliefs. The Author can only present the physical workings of the Doppler Earth and the possibilities. Doppler space-time enables us to view existence beyond our Earthly lives. It widens our perspective of the world of the living and the world of the dead.

SECTION 16-2 ROTATIONAL SPIRITUAL PHYSICS

The large spiritual energy comes from the Earth's orbital motion. This produces a Doppler Earth and a huge space-time memory system encompassing the entire orbital ellipsoid of the Earth. The rotational Doppler spiritual energy of the Earth is less because we only move at approximately 1000 miles per hour on the surface of the Earth. The rotational energy of the Earth as compared to the Orbital energy is very small. This is especially true since it is the square of the velocity that counts. There are approximately $66^2 = 4356$ times the amount of Einsteinian spiritual energy external to the Earth as part of the Earth.

In spite of the numbers, there still is a huge amount of spiritual energy within the Earth itself. To understand this let us look at a radio tower at the equator. The tower is sending out low frequency radio waves, which travel all around the Earth. If we look at a receiving station on the opposite side of the Earth we find that the frequencies coming from the tower appear to be the same on both sides of the Earth. Yet the Earth has moved. If we put up a perfectly stationary satellite above the Earth, which is moving relative to the Earth's surface, we would notice different frequencies coming from opposite directions

From this simple experiment, we find that there are two Earth's. There is the ordinary Earth and there is a Doppler image of the ordinary Earth. If we look at a point on the Earth we find that there is a Doppler image of that point on the opposite side of the Earth similar to the Doppler Earth on the opposite side of the sun.

If we spin the Earth very rapidly we find that the Doppler image moves closer to the original point. If we spin the Earth at the speed of light, the Doppler image would be at the same exact point. In effect we would have a tremendous amount of energy built up within the Doppler Earth or the Ghost Earth. We get similar effects in the cyclotron.

We see that energy above and beyond ordinary rotational energy causes a ghost image of the Earth. This is the spiritual Earth. Since the Earth spins slowly, the Ghost Earth is a very low energy Earth. Yet, this energy permits the memory of yesterday to be stored within the sphere of this Earth. We then have two levels of a spiritual space-time soul. We have the rotational space-time soul and the orbital space-time soul.

Without both of these motions, life as we know it would not exist. We have two spiritual Doppler forces acting upon us. We have an orbital Doppler that produces differences between front and back. We also have a rotational Doppler force, which causes differences between left and right. These two spiritual Doppler forces produce creatures with fronts and backs, and lefts and rights.

If we found a planet that only orbited a sun, then the life forms would be perfectly symmetrical left and right. They would only have front and back differences. Therefore man cannot be produced on a planet which does not revolve on its axis.

To make matters worse, if the Earth's orbital rotational speed dropped to zero, then the life that we know would perish. The moon does not rotate on its axis with respect to us. It has a face lock condition. Yet it still rotates as part of the entire Earth/Moon system. We should be able to survive on the Moon for a long time. We may even be able to produce children on the Moon. Yet, we must carefully study the long term effects of us not being able to have the same axial rotation as we have here.

The same is true for the other planets. We may be okay on some. However our ability to reproduce may start to die out if we stay too long on a particular planet without sufficient axial rotation. These are things to be studied in the future.

Some of our space-time soul is stored in the rotational energy of the Earth. This will discharge into the orbital motion of the Earth. This in turn will flow into the Doppler Earth. There it will be combined with all others and head back toward this Earth. It will enter the spiritual Earth and then flow into life. We have a complex space-time machine which keeps our soul active in death and then processes our soul.

SECTION 16-3 PRODUCTION OF ONE CELL LIFE

Let us look at the production of the first one cell life. The first simple one cell life originated many universes ago. The Mind of the Universe remembers the structure of one cell life from the previous universe. Therefore we merely repeat a process which occurred before.

Let us now go back several hundred big bangs ago when Earth's started to be produced by the gravitational pump. The production of Earth's with water is necessary for life, as we know it. When we look at basic life we look at rock crystals. Crystals grow by repeating their chemical structures. Dripping water such as at Howe Caverns in New York State cause interesting structures to form. These are living rock structures. They are intelligent structures and are similar to organic life.

The basic rock structures have spiritual dimensions and reflections upon the Doppler Earth. As we look in the direction of travel for the rock structures we see a higher Doppler frequency in front of the structure in the direction of the motion of the Earth, and a lower Doppler frequency behind the structure. This is due to the Earth's orbit. A spiritual image of the rock structure exists in space and time and with a foci point upon the Doppler Earth.

Our Earth rotates; therefore it contains a rotational Doppler as well. The rock structure has a rotational Doppler on the opposite side of this Earth as well. Since the Earth only rotates at 1000 miles per hour, this rotational Einsteinian Doppler energy is much less that the orbital Einsteinian Doppler energy, since the Earth is moving at 66,000 miles per hour. However this does not matter. The Earth's orbit produces complex revolving standing wave patterns of the Earth's rotation.

The net effect is that we have a soul of the rock formation in orbital energy and another soul of the rock formation in rotational energy. We also have an expanding Earth. This produces a third spiritual image of the rock as the Earth expands.

The result is that the rock crystal has electrical waves which form complex patterns external to the rock itself. In fact when we look at the outward motion of the entire universe and the counter effect of gravity, we find that a simple rock has extremely complex standing wave patterns which encompass the entire universe.

The net result is that the simple rock structure has external forces acting upon it, which cause it to live and grow. The rock grows due to the Einsteinian energy and the Doppler effect. These forces are called spiritual energy because they are the driving forces acting upon the material rock.

The spiritual forces are really electromagnetic space-time forces. They are not things that we do not understand but the word "spiritual" aids and assists us in understanding the complex electrical process. It is the real space-time forces, which act upon matter. We then have a world of matter and a world of spiritual forces, which are the space-time forces due to the rotation of the Earth on its axis, the orbit of the Earth around the sun, and the expansion of the universe which produces a back-pressure or gravity.

The spiritual forces are complex. They involve standing wave patterns. They involve pressure exerted upon matter to move atoms and molecules to match spiritual images. The spiritual forces are complex space-time electrical forces, which operate upon matter to produce life. The crystals we see are basic life. There is a Life-force of the crystal. The development of a single crystal upon the face of a primordial first Earth caused a primordial Life-force of that crystal to be born. Random chance produced a stable pattern. This in turn caused the image of that pattern to occur upon the Doppler Earth. This feeds back all over this earth. The net result was Howe caverns everywhere the same conditions occurred.

The living cell that we are familiar with is an organic cell. Therefore it is a much higher complexity than the simple crystal cell. The main ingredient for life is water. Water permits the molecules to oscillate freely. Let us look at a primordial universe long ago. We see the ingredients for life upon the seas. By random chance several ingredients came together. They formed a local pattern. Then they formed a space-time pattern over hundreds of millions of miles. Let us assume the Mind of the Universe was devoid of any memory at that time, and that the universe sat in primordial ignorance.

The water permits the ingredients to oscillate readily at the large scale level of molecular size. This produces the added dimension of space-time chemical oscillations. The ingredients had a little heart beat. The net result, when we look in the spiritual space-time dimension, is a very complex electrical wave-shape, which produces perfect standing waves in complete harmony. We then have stable life. In addition the little one celled creature absorbs similar atoms all around it. It grows. We now have a spiritually stable configuration of life produced by pure random chance.

Once the first cell is accidentally produced and is stable both locally and spiritually, we have a living cell, which becomes its own simple Life-force. Upon the Doppler earth an image of this one cell Life-force emerges. This will feed back upon the material Earth and everywhere similar conditions exist, a force will occur which moves the chemical soup into the formation of the one cell creatures.

These are one cell vegetables and they will grow all over the Earth. They will have a chemical heartbeat since they will oscillate. Ordinary crystals only oscillate small distances. Complex organic crystals have a much wider oscillation.

As we look at the one cell critters we see they have a longer Doppler length in front of them in the direction of orbital motion, and a shorter Doppler length to the rear of them. They have a slightly different front and back. They would be completely symmetrical due to Earth's orbit alone but since the Earth also rotates we have a difference between left and right. There is Doppler rotational force which produce left handed cells. Thus all life upon this Earth will have left handed cells due to the Earth's rotation and the Earth's orbital motion. If a different Earth rotates the opposite direction, the people produced will have right handed cells.

We see that basic life has a heart beat and eats. It also has left handed cells and is not perfectly symmetrical. It has a front and back. We see that pure electromagnetic field theory helps explain the formation of early life.

The cells will grow. An optimal size will be reached and they will split. Structures of many cells will form. In addition many different type of basic cells will be produced. For every step of the way, the projections of the standing waves far out into space must be stable.

One set of cells adds to another set of cells. The higher configuration produces a Doppler image upon the Doppler Earth and a Life-force forms. This Life-force produces pressure, which causes more combinations of the same set of cells to form. Every time we move upward in complexity we move from basic primordial intelligence toward the intelligence of grass and weeds.

The primordial cell moves upward and becomes the heterogeneous vegetable Life-force. It moves toward higher and higher combinations of these cells. By the time we reach a tree, we are achieving a very high degree of vegetable intelligence. The vegetable Life-force is not at our level but is quite smart in its own way.

One cell animal life formed early to feed upon vegetable life. The process for the production of higher animals merely is the continuous additions of complex combinations of cells. We move up higher and finally reach the level of man.

Man was not produced upon the first primordial universe. We are the result of many prior universes. The initial simple process for the production of life was modified by remembered wave-shapes of the original one cell creatures. It took many universes before the eye was perfected.

The intelligence of the universe slowly builds up. It was homogeneous stupidity long ago. It oscillated and grew a rat. Later on man came upon the scene. Man came after the rat. The process remains the same however. The advent of man caused the spiritual world to move from a mere memory device and focal point to a real Human Life-force.

SECTION 16-4 THE STEPS OF EVOLUTION

Let us look at a chart of the evolutionary steps toward the production of man and higher man.

EMF
Pre Universe
Universe
Galaxy
Solar System
Planet Earth
Pre Life
Early Life
The Dinosaurs
Modern Life
Chimp/Apes
Pre-Man
Man
Higher Man

As we look at the chart, we see that the universe started as an electromagnetic field (EMF). This field became the pre-universe, which oscillated from big bang to little bang hundreds of times until the present universe was produced. The production of the universe gave birth to the galaxy, which gave birth to the solar system and finally the planet. It took many universes and many evolutionary steps before an Earth was produced.

The production of an Earth gave birth to pre-life, which gave birth to early life, which produced the dinosaurs. Then for later Earth's, modern life evolved, and the Chimp/Apes formed. The Chimp/Apes lead to Pre-man, and finally man. Higher man is the final step in the evolutionary process.

In this chart we see that a lot of photographic memory steps drove the evolutionary process for the production of man. An infinity of time occurred before the first crude man on the first viable universe. From there it was still an infinity of time before man achieved what we see today. For every universe the steps are reproduced over and over again. For better or worse, man wants to live again and again.

Man could survive indefinitely upon the Earth in small numbers. However man has no population control feedback chemistry built in. Higher man has higher spiritual chemistry built in which limits population. The same is true of the animals for the next creation. They will slowly all turn vegetarians so higher man and higher animal life will be a superior existence. Man at present overpopulates and uses up the resources of the Earth. This results in the destruction of the ability of the Earth to maintain both animal and vegetable life. In the end man fights man for food and water until man self-destructs. The destruction of man produces tremendous spiritual energy, which causes higher man to be born to the few remaining survivors.

The survival of the species depends upon man maintaining a selected few of the species in outer-space as man and the Earth self-destructs. The few who return to the destroyed Earth will find the evolutionary process working well and cleansing the Earth for their offspring. The means for the survival of man in space is illustrated starting in Chapter 17. It is our moral necessity to provide the means for the survival of man.

The production of man changes the usual collective reincarnation/evolution process into a greater complexity. The whole reincarnation process and the purification of the Collective are a higher evolved spiritual process. This higher process leads to higher man and a future where the physical life of a whole reincarnate moves into a spiritual life within the highest evolved level of the Human Doppler Mind.

In this manner life and death are continuous, with the person being maintained. Such things are for the future as higher man comes of age. We still exist at a low level of spiritual development. Most of us merely become collective reincarnates in which our individuality is lost to the collective in death. Only some of us become whole reincarnates. Even less of us achieve the inner portion of the Collective, which brings us into future existence. All of this is made possible by the space-time memory properties of the universe and the Doppler Earth. All this makes the world of the dead both possible and quite interesting.

SECTION 16-5 THE FINAL PHYSICAL MAN AND WOMAN

Our Earth and other Earth's throughout the Universe will undergo many upheavals and transformations over the years. Higher man will last a long time. Higher man will eventually live a nearly perpetual life.

Higher man will eventually turn into highest man. Highest man will be immortal. He will live with a near perfect body. In death he will enter the purified inner mind of the Collective. He will then await his rebirth. Every man born will be born again as himself perpetually. Death then becomes meaningless. Of course everything must come to an end.

Eventually, the window of life will start to fail in the universe. Stars will explode. Many Earth's will be destroyed completely. Some Earth's will survive. Yet as time goes by the populations of highest man will decrease.

Children will cease to be born. The population of highest man will slowly decline. Highest man will live perhaps a billion years at various locations all over the universe. He will slowly evolve to live longer and longer. His body will attain immense perfection. The same will be true of the entire creation.

This has been the rough level of existence. We are evolved existence. We come from primordial intelligence, which worked its way up to man. The collective intelligence was not there. Higher man is the creation of us. It is man perfecting himself. Higher man contains all the intelligence and medicine we generated. Higher man stands at a higher physical plane.

Higher man also stands at a higher moral plane as well. Higher man knows where it came from. Higher man knows the horror of our existence. Higher man collectively strives for greater humanity and perfection. There is no barrier between higher man and the Life-force. They are all one and the same. We have achieved as best as we can. We are lower men with too many diverse viewpoints. We are still tribal men where every tribe stands against every other tribe. We stand separated from the totality of the Collective. Higher man starts as tribal man but where everyone is a member of the tribe and no one is separated from the Life-force.

The important thing is that higher man will insure that every member stays a member of the tribe. They will treat everyone as their brother. Higher man tends to be idealistic. Higher man is also the purified collective of lower man. Thus higher man starts life with a greater purity and purpose.

The Human Collective undergoes the greatest pain and suffering when all of mankind is destroyed. The individuals suffer little but the Human Collective suffers greatly. Higher man is a product of that intense suffering of the Human Collective. It takes a lot of pain and suffering on the part of the collective of mankind to produce a product which has immense humanity and humility.

The population of higher man will be limited to several million people. The souls of billions of people will be purified and in the end, only several million people will be produced. The people who are born of this process are rich of soul. They have a depth and intensity equal to hundreds of people. The majority of mankind will revert to the lower animal population. They will be superior loving animals, which do not eat each other.

As time goes by into the far future, the energy of the Earth will become depleted. The population of the Earth will start to shrink. The ability of the Earth and all other Earth's in the universe to support life will decline. Each higher man will have lived thousands of years at the peak of higher man's life. This eventually will start to shrink.

We reach a maximum state of perfection and then things go down hill. As many people die out, their spiritual energy returns to the Life-force. This in turn flows into the remaining people. The few then live for the many. Then a final decline starts.

Fewer and fewer children will be born. People will start to die younger and younger. The body of man will achieve maximum perfection and then start to decay rapidly. Once maximum perfection is achieved the process tends to self-destruct.

In the end all the spiritual energy from the Life-force will be flowing into only one man and one woman. It will be Man and Woman in a final Earthly Paradise. It is Man and Woman. It is the Human Collective Itself. It is the final end of both man and the Collective. It is sad. Two beautiful lovers are all that remains of mankind upon this Earth.

The lovers will slowly die together in each other's arms. The entire paradise of the New Earth will also die out. In the end, we will have a dead Earth. The two lovers will die frozen in each other's arms. They will be naked. The Earth will freeze. Their bodies will stay until the end of time.

There will be no bugs to eat them. The bugs will just be frozen like everything else. The Earth will have moved outward away from the sun and the sun will have perished as well. In the end we have the remains of Man and Woman in a final embrace for all eternity.

Then the universe ends. The ancient story of Man and Woman in Paradise will return. Man and Woman started as simple creatures of the Life-force. In the end, they will be super intelligent Super-Man and Super-Woman. The end of Paradise will be fast forwarded to the next universe. Man and Woman will be remembered by the Mind of the Universe and we will evolve again.

SECTION 16-6 MAN AND WOMAN IN A SPIRITUAL PARADISE

Our Universe will end with Man and Woman in an eternal embrace of death. The little bang will occur and all the protons, electron, neutrons, and photons will be converted to electro-dot energy. The bodies of Man and Woman will be destroyed. The universe will stop expanding and start contracting toward the next big bang.

The next universe will begin. We will again find an innocent Man and Woman in a Garden paradise. Things will be a little better next time around. Yet the story always starts with Man and Woman in a physical paradise and ends in the same state.

The universe is an electrical oscillator, which cycles toward greater and greater perfection. The universe will oscillate with smaller and smaller ratios of maximum radius to minimum radius. When the maximum to minimum ratio approaches unity, the big bang and the little bang occur simultaneously.

Under this condition, no electrons, protons, neutrons, or even photons are produced. All we have is electro-dot energy. All we get is a singular space-time mind filled with the knowledge of all things. Near the final end of this process, we get a singularity of the Mind of the Universe in active intelligence.

This Mind will produce Man and Woman in a spiritual paradise in pure electro-dot energy. Man and Woman can travel the spiritual universe at the speed of light. They can live forever. Near the very end of the Universe a magnificent Man and Woman in a spiritual paradise can be born.

For each cycle, the lovers died in each other's arms. They died into cold death. They were the utmost physical perfection. They were the final man and woman. Yet in the end of the process, out of the death of the two lovers arise the magnificence of a spiritual paradise.

The Mind of the Universe is but a machine. However the machine appears to be able to become a being. The Life-force of the Universe becomes us. The Life-force is a machine who thinks like man. The process is complex. There are a lot of Darwinian steps along the way.

It is a beautiful process. We start with a simple innocent Man and Woman in a garden. We rise upward to man. We end up with the death of the physical Man and Woman. Finally we find a spiritual Man and Woman in a perpetual paradise.

Man and Woman in a spiritual paradise will exist in that state for an infinity of time. We could say that they would exist forever in the final state. In truth however, the electrical laws must be obeyed. The oscillation rate of the universe will become slower and slower. Man and Woman will slowly grow old.

Every thing, which could have been done by man, will have been done. As the universe cycles ever slower, Man and Woman in Paradise will slowly go to sleep. The universe will reach a steady state. Living spiritual energy will become pure standing waves. The living Man and Woman will become a beautiful frozen spiritual sculpture. There will be no thought in the universe. The universe will have reached a state of absolute perfection and then it will rest in eternal quiet.

Perhaps the final spiritual sleep or death of Man and Woman and the Life-force will last forever. Most likely, the last artwork will self-destruct and a new Universe will be born. There is an infinity of universes possible. This universe may never repeat again. We may reach the state of absolute perfection and the Life-force's work is complete. In the end, the Life-force created the most magnificent living sculpture.

We suffer a little. We suffer so that the final perfection will be accomplished. Some of us will take part in the final creation. It will take billions of men to produce the final Man. It will take billions of women to produce the final Woman. All of us can strive to move ourselves upward toward greater love and humanity so that we can continue in the process. Each of us has the choice to strive for perfection or to strive for annihilation.

PHYSICS & SPACE TRAVEL

CHAPTER 17: PREPARATION FOR THE TRIP TO ALPHA CENTAURI

SECTION 17-0 INTRODUCTION

In this chapter we will look at what is necessary in order to travel outside the solar system toward Alpha Centauri. We will find how far and how fast we can travel into deep space before we develop physical problems. Then we will have to turn back. In the future we must be prepared to survive in space, on the moon, and the outer planets for up to forty years while the Earth is uninhabitable due to atomic radiation, intense solar flares, turbulent volcanoes and earthquakes, and or comet strikes. We must also learn to move from star to star in a quest for viable Earths.

SECTION 17-1 THE PROTON THRUSTER ENGINE

The proton thruster engine is the most important physical device for the trip. We must be able to travel up to fifty percent of light speed (0.5C). In addition we must maintain continuous acceleration at 32 feet per second per second (1G). Weightlessness will not be acceptable except for very short periods of time in terms of minutes.

The last requirement causes additional power requirements on the engine. We must accelerate uniformly at our rate of gravity of 1G, and reach near one half light speed. This will take approximately 6 months. Then we must decelerate for periods of two weeks at a time. This will reverse the gravity of the spaceship from bottom to top. The passenger compartment must be built as a mirror image or be readily reverse-able. Then we will accelerate again for two weeks.

The ship will have two main proton thruster engines. The rear one will bring us up to one-half light speed. The front one will protect us against anything in our path and produce the reverse gravity. When we slow we will increase the front thruster and decrease the rear thruster to near zero. When we accelerate we will increase the rear thruster but maintain a slight protective front thruster.

As we travel near half light speed, we will produce gravitational waves in front and to the rear of us. These will act as protective barriers as well. The front of the spaceship must come to a point, which provides a protective Doppler image far in front of it. The gravitational waves, the Doppler image of the point, and the front thruster will permit us to travel safely.

If we find a large rock ahead of us, we must increase the front thruster and the rear thruster in synchronization. The power will increase tremendously but the crew will still experience a force of only one G. The front thruster will completely destroy the rock and everything will be safe.

Of course we cannot destroy a huge space rock. Therefore we must always look ahead and see what is there. If a space rock is reflecting sunlight, then we can see it. If a dark huge rock is in our path, we will not see it. Our front thruster will rebound from it and our computers will have to decide what evasive action is necessary. In general we will have severe problems only in certain areas of space. We must chart all space debris carefully before the trip. When we reach near Alpha Centauri we will slow down to less than 0.1C. At this slow speed we can take evasive action readily.

The heart of the spaceship is the proton thruster engine. The proton thruster engine converts protons into pure photon energy. It is a controlled proton bomb except that only several protons at a time are broken apart.

Protons contain only dots within them. They are concentrated electro-photons. The easiest source of protons will be in ordinary water. The fuel of the proton thruster engine will be pure water. It is a fancy electromagnetic laser type engine. We cannot achieve the proton oscillation frequency but we can achieve a sub-harmonic of this frequency. By pumping out square waves at the sub-harmonic and simultaneously producing strong electromagnetic and electrostatic laser fields, we will rip apart the proton. These engines most likely are in common use all over the universe by advanced man.

It is only a matter of time and money before we can produce the proton thruster engine. It will probably take between twenty and fifty years to produce the engine. The proton thruster engine can be made small. It can be used for good or for evil

The output of the proton thruster engine is pure photon energy. It is a beam of light and X-rays. It is dangerous. It can be used to destroy missiles aimed at a country. It can be used for protection by governments. It can also be used to destroy enemy aircraft, satellites, and ships.

The proton thruster engine can be a weapon of war. It is also the means by which man shall ride the light and travel the stars. An International Alliance will develop it. Several governments will become involved. The theory is simple. Break apart the proton and obtain its energy. The physical construction required to do this will take several billions of dollars and hundreds of engineers and scientists to do the job.

Without the proton thruster engine we have to used old-fashioned methods of huge booster rockets. It does not seem likely that we can achieve Alpha Centauri and beyond without the proton thruster engine. It may take the wealth and power of the United States and a mixture of other countries to fully develop the engine. They will work to advance the prototype design as far as it can go for a good working model. Then some large corporations will have to start the manufacture and mass production of the engine.

Once the engine is developed and all countries agree not to use it for harm, then all countries can start to build their own spacecraft. Some will use it to go to Mars. A trip to Mars will be fast and easy. Others will explore the solar system. The initial use of the proton thruster engine will be to travel our local solar system.

The engine permits high-speed space travel up to 0.5C without the necessity of weightlessness. Ordinary people can ride the spacecraft. Ordinary people can travel to Mars. Takeoffs and landings are easy. It is ride the beam up, and ride the beam down.

The proton thruster engine permits easy space travel. A frail elderly man could travel to Mars. People can be born on the spacecraft. You can leave the Earth at 1.5 G's or less. A man who weighs 160 pounds will weight 240 pounds during takeoff. Once you leave the Earth, you will only weigh 160 pounds. The frail elderly person who wants to travel to the moon can take off at only 1.1G's. A 120-pound woman will only gain 12 pounds on takeoff.

Modern rockets are tough on the individual. The proton thruster engine permits one to be in an elevator in the Empire State building. It permits absolute control over the acceleration of a huge spacecraft or even a small spacecraft.

The danger of the engine is that nothing shall be behind the beam. When you take off, you will start to drill a hole in the ground below you. You will melt and vaporize everything. This is great in oil well drilling. You can use a proton thruster engine to drill through rock. It will produce molten rock on all sides. You can make a rock pipe of any desired diameter you want. You can drill fifty miles down if you want to.

It is necessary to build the proton thruster engine. This engine will bring us readily to the outer planets and possibly to the stars. The future is bleak and we must move on. Man will move toward higher humanity and self-preservation and use the future technology to fulfill our destiny. We must travel the stars and we must not use the proton thruster engine for harmful purposes.

SECTION 17-2 THE NON-RADIOACTIVE ATOMIC GENERATOR

Another device of great importance for the space ship and space stations is the non-radioactive atomic generator. We need clean safe electric power to heat and run the spacecraft. On the Earth today we use coal, oil, natural gas, water, and wind as sources of electrical power. We also use radioactive substances such as uranium. The use of radioactive substances produces danger to mankind during their production. It also poses danger to mankind due to accidents. In addition, it produces harmful waste products.

The Proton thruster engine design can be used for the direct production of electricity. The fuel will be water again. The protons and electrons will be separated and the photon energy will be used to produce a DC electric current flow, which will run DC motors or DC heaters. The DC motors will run AC generators for the distribution of power over large distances. The non-radioactive atomic generator is a modification of the proton thruster engine. It produces electrical power instead of rocket thrust.

These are the generators of the future upon the Earth. Oil will run out. Coal reserves will be depleted. However, only water is necessary for the non-radioactive atomic generator. In the future small versions of the generators will be produced to run cars and trains as well. We must protect against X-rays. This is no problem in electrical power plants or for propeller airplanes since shielding is readily accomplished.

We will return to propeller airplanes driven by the DC atomic generators. They will be safe. The proton thruster rocket is not safe in general. It might be possible to run one underground train using the proton thruster engine. However nothing can be behind the beam. Yet if we run one train at a time we possibly could use the engine using extreme safety precautions.

The non-radioactive atomic generator does not have that problem. All we are producing is electrical power. With proper design, only some ordinary heat will be produced. In the future clean cheap electrical power will be produced using proton thruster techniques adapted to ordinary electrical power.

These power plants will not produce smog. The trains will run clean. Busses will run clean. The automobile may have to use battery power if a small enough generator cannot be produced with sufficient shielding. It is merely an engineering problem. The smaller we make the engine, the harder it is to shield it. However, the power level is much smaller too. It may be possible to bring the engine down in size for use in an ordinary car.

Upon the New Earth private cars are not in the best interest of society. Trains, busses, airplanes, etc. are used. Everything is community property. The automobile does not fit into a collective society where every-man is the brother of every other man. In addition, the automobile is quite dangerous. Therefore it is not a product destined for higher man.

In the future upon this Earth, every apartment house can have its own generator plant. Since the fuel is ordinary water, there is no necessity for great power companies. All you need are service companies, which adjust the generators and maintain them. With proper design service should be about once a year.

We see that in the future, smog and pollution will ease up as we replace coal and oil with water. On the negative side we are still using a great amount of heat energy. However, the non-radioactive atomic generator if it existed today would help to reverse the greenhouse effect. With determined birth control in the future, it might be possible for mankind to live here for quite a long time. Yet, we eventually will self-destruct since we are people who are always in conflict.

Upon the New Earth, the non-radioactive atomic generator will permit man to live there for an indefinite amount of time will little environmental damage. In addition higher man upon the new Earth will have a small population of from one million to ten million. Their civilization can live one million years or more at a very high standard of living.

In effect, as compared to here, every man upon the new Earth will be extremely rich. Food, housing, power, recreation, etc. will be available in great abundance. Medical care will be top notch as well. Once we arrive upon the New Earth we will move upward in humanity, health, and wealth.

SECTION 17-3 THE ABSOLUTE VELOCITY SENSOR

Between the proton thruster and the non-radioactive atomic generator, we have the two main ingredients for the space flight to the outer planets and to Alpha Centauri. It will also be necessary to build some better control instruments.

On the trip to Alpha Centauri we can measure the velocity of the spaceship by observing the color of the light coming from both Alpha Centauri and our own sun. The faster we move the bluer will be the light from Alpha Centauri and the redder will be the light from our sun. We can then build an instrument, which measures the light from both stars and produces a velocity measurement. It will tell us our true velocity to a very high degree of accuracy since both our sun and Alpha Centauri are moving slowly with respect to the speed of light.

Once we are far enough from our sun and far enough from Alpha Centauri we are no longer effected by either stars gravitational field. We can then use an optical instrument similar to that used during the Michelson/Morley experiment. In pure free space we are no longer part of the Earth's gravitational field or the Sun's overall gravitational field.

The test instrument, which produced a null upon the Earth, will find itself producing patterns of distortion upon the spaceship. As the velocities move upward, the distortions will become the means by which absolute velocity can be measured.

The scientists aboard will have plenty of time to calibrate the instrument. Most likely it will be calibrated during initial trial trips to Mars and Venus. The final instrument will not only pick up the relative motion of the spaceship but the absolute motion as well including the galaxy motion.

The motion of the Earth and the Sun tends to produce common mode distortions upon the Earth and all the Earth's instruments. Since the Earth is in a basic steady state orbital pattern, we cannot readily produce absolute motion detection. In addition the mass of the instrument is tiny compared to the Earth.

Once we are in a spaceship in deeper space, the mass of the instrument is only compared to the mass of the spaceship. In fact we can produce the instrument as part of the entire spaceship. Then the instrument stands independent of any external mass. This enables accurate readings of the absolute velocity of the spaceship.

We can always rely upon the Doppler readings of the stars for our velocity. However it is always useful to have two means of velocity measurements. If we move too fast we can hurt the crew or damage the spaceship. In addition we need to know the exact speed where the crew and passengers start to feel uncomfortable.

Somewhere between 0.4C and 0.5C problems will arise. We need to record the velocity where people start to feel ill. Then we must set the computers to prevent exceeding this velocity. If we go too fast we will harm everyone.

The absolutely velocity sensor with computer readouts all over the ship will provide all with an understanding of our speed. Also the ship must contain acceleration readouts so that people can relate to the effects of various accelerations. If 0.25G is okay after we have achieved maximum speed, then we can use this number. However, over a long period of time, people will start to feel ill at various low values of acceleration. Both velocity and acceleration must be monitored. The acceleration monitors can be simple devices since accelerations will be similar to that of an elevator. Masses and springs may suffice to do the job even though more fancy electronic ones will most likely be used. However more than one type should be available with the simple mass/spring mechanism as backup.

SECTION 17-4: SPACESHIP REQUIREMENTS

In this section let us look at some of the requirements for the spaceship. As we look at some of the modern aircraft carriers we find them to be approximately 1000 feet long. They are approximately the size of the Empire State Building in N.Y. They have a crew of up to 5000 people.

We need to build a spaceship, which will carry up to 250 people to Alpha Centauri and beyond. We need to have enough supplies of food and water to last at least ten years. The spacecraft will have to have trial runs to nearby planets. People will be able to travel to Mars for study and tests. We will build various space stations upon the local planets where possible.

We will travel to all the local planets for scientific purposes and to gain experience in the use of the proton thruster engine and to perfect the spaceship design. We will also find out if man can survive for long periods in deep space. The first spaceships may only carry 50 or 100 people to the local planets. However, on the trip to Alpha Centauri we must carry at least 250 or more people.

A mother ship will need perhaps two manned or unmanned supply ships on the trip to Alpha Centauri. We could send out the supply ships first or travel as a fleet. The food is not going to be very pleasant. We will have to develop good tasting quick foods. Most will be freeze dried or powdered. However, hopefully we will develop reasonably pleasant food substitutes. Some canned goods will be okay for the trip. We do have excellent refrigeration aboard the spaceship.

There will be plenty of heat and hot water. Everything will be recycled. Microwave type cooking will be used extensively. The amount of food and other supplies will determine the size of the spaceship and the number of passenger plus the number of cargo ships required. Columbus sailed with 3 ships and we may need 3 ships as well.

By the time we have traveled to the local planets, we should have gained great experience in how the people react to the spaceship. We will never achieve 0.5C on our small trips to the local planets. We will have to go for a trial run of six months to achieve maximum speed.

At 0.5C the forward Doppler length of our bodies is almost twice our height. This will cause us some problems. The spacecraft ceiling of each floor will have to be at least 20 feet tall. If we built a spacecraft of 500 feet, then it would have 25 separate levels of 20 feet each.

The exact maximum speed we can travel is limited by the point where we start to feel ill and are unable to function well due to our increased size in the direction of travel. We shrink gravitationally but enlarge inertially as the velocity increases. This causes distortions in our bodies and a degree of distress. During the long test trip careful studies will have to be done to find the maximum velocity we can achieve without long-term harmful effects. Hopefully we can achieve between 0.4C and 0.5C without discomfort.

The ship must be like a cruise line of sorts. There must be music and entertainment. Everyone must have a job. We will have to be at least ten years aboard the craft. Therefore each day must be filled with activities. We must mix work and studies. Music is very important on the trip. We need doctors and comedians as well.

People must be kept busy and interested. Education is a big plus. Everyone will be trained in several jobs. The plumber may be a cook and a musician as well. The doctor may be a teacher and an astronomer as well.

The ship must carry a library in paper and also on the computers. When people arrive at their final destination, most will all be fisherman and farmers for awhile. During the long trip all people will be taught how to farm collectively. The trip will be a continuous effort to impart survival skills to the 250 or more people aboard the ship so that they will be ready to reestablish the New Earth on the return trip.

The spaceship must carry a good emergency medical facility. Unfortunately it will not be possible to maintain the same level of treatment as here. Some people will die on the trip. The library must both be in paper and also in electronics. A basic minimum must be in paper to insure the recreation of our level of existence rapidly. Much will be stored on computer disk but in the event of damage, this will be lost until the new planet is ready to build new computers. The plans for the computers must be on paper.

In the event of problems, it might take fifty or one hundred years to bring us to our present technical level upon the New Earth. We must be prepared to temporarily move backwards in time to days long ago. It will be like camping out without electricity for awhile. However we will always stay near the spaceship for medical care and electricity. It will be the center of the new city.

The spaceship should provide a modern city center for the New Earth. Some people should be able to live there for quite a long time. A city then could be built around it. However, the spaceship must land on solid rock and weld itself to the rock. An earthquake or other local weather problem could damage or destroy the spacecraft. Man could revert to being upon a new planet without his state of the art abilities. The initial struggle to survive necessitates farming and fishing skills.

The spaceship must come to a strong physical point with a power thruster within the center of this point. The point must be very strong tungsten steel or the like. When we travel near half the speed of light, the tip of the spacecraft will shrink gravitationally but extend inertially. The tip will appear in softer energy In front of the spaceship. This will protect the spacecraft against small rocks and various particles in space. The inertial tip will produce an extended inertial wave-front in front of the spaceship. The front proton thruster must always be kept on at low power as well to destroy anything in the immediate pathway of the spacecraft. Otherwise it would not survive the impact of any small space rocks. If a larger rock appears, then the front thruster must automatically increase power while simultaneously the rear thruster must increase power as well. This will destroy the rock.

Continuous radar must send out signals in front of the spacecraft and look for rebound patterns so that the front thruster power can be increased as necessary. A point is reached in ship speed where electronic systems deteriorate.

Einsteinian space-time was incorrect for linear motion at high velocities. He felt that each system was independent of velocity but in truth as soon as we achieve close to 50% of light speed, we will experience some problems with velocity. The radar system will start to fail. The antenna systems will start to distort terribly. A point is reached where the various systems and the computers as well start to fall apart. The exact speed at which the spacecraft can travel will be found on the early trips to the planets and beyond the planets as a good trial run is made to achieve maximum speed. It takes six months to achieve 0.5C so the trial trip would take over one year.

CHAPTER 18: TRIPS AROUND THE SOLAR SYSTEM

SECTION 18-0 INTRODUCTION

In this chapter we will explore the initial trials and tests of the spaceship destined for the trip to Alpha Centauri. We will also explain the requirements for the space stations on the Moon, Mars, Venus, and Pluto.

SECTION 18-1 INITIAL TRIAL RUNS AND TESTS

As soon as the proton thruster engine is completed, trial runs can begin with small versions of the spaceship. Since there is no explosive fuel aboard the spacecraft, we do not have to worry about fire and explosions aboard ship. Of course electrical fires are always possible. However ground fault detection circuitry and other electrical protective circuits can readily prevent these problems.

The non-radioactive atomic generators will supply all the electrical power necessary. The generator will have been in use upon the Earth for a long time prior to the trip. It will not be necessary to prove it out. The proton thruster engine will also have been used in defense and for oil mining upon the Earth as well. Therefore the main power sources will be proven devices before the first trial runs of the spacecraft.

The fuel is water and as the water is used up and converted into light, a vacuum will result in the storage holds of the spaceship. The storage tanks must be similar to the tanks used by natural gas companies, which expand when full and contract when depleted. In this way, the depleted tanks are open to the vacuum. An alternative is to use up the many small tanks and then expose their insides to vacuum. This will be part of the mechanical design of the spacecraft, which must be carefully tested during trial runs.

The proton thruster engine rides the light beam. It will start to burn a hole in the ground directly beneath it. Since the spaceship accelerates very slowly, it is more like an elevator than a rocket. For stability the front thruster will be used initially as well. The spacecraft will lift off slowly in a perfectly vertical direction.

The beams must be modulated so that the craft will follow the spin of the planet. We must keep the spacecraft in a perfect straight line. Therefore the motion of the atmosphere must move the spacecraft in a constant position orbit. After awhile, the spacecraft will be moving at a large horizontal speed equivalent to the Earth's rotation. The aerospace engineers and scientists will work out the details of this.

If we kept the beam on, then we would be fighting the wind for the beam will keep us on a perfectly straight line whereas the Earth is moving. We either need to have some small sideways thrusters or we let the atmosphere move us while we turn our main thrusters off and on as needed. An ordinary rocket pumps out gasses, which react with the atmosphere. This gives the rocket a sideways force. The proton thruster light beam basically becomes its own reference. It tends to move the rocket in a perfect straight line. The result will be that the Earth will spin under the rocket. Therefore we must turn the engines on and off to move with the Earth's rotation. Otherwise the beam will start to damage the land underneath the rocket.

Once we leave the Earth's atmosphere, then we can slowly turn the spacecraft by small perpendicular thrusters toward the direction we want to go. Alternatively we can have the main thruster consist of many different sections on a plane surface. By increasing one section over another, we will turn the spacecraft as desired. This will all be part of the spacecraft design, which will take several years to accomplish by experts in the field of rocket type design. This is the job of aerospace and mechanical Engineers. They will have to learn about laser rockets however.

It takes six months to reach 0.5C at a 1.0G rate of acceleration. We can reach 0.1C in about one month at 1.2G acceleration. The initial long-term test will involve one month of acceleration and one month of deceleration. We will end up at zero speed in deep space. Finally we will slowly turn around and return in two months. During this long trip we can experiment with how much acceleration we feel comfortable with over long periods of time. This will test out how the different levels of gravity effect us.

We have become familiar with long term weightlessness. On the trip we will learn the long-term effects of various levels of acceleration. The trip also gives us a chance to test out the spacecraft and learn of any deficiencies. Some corrective actions can be taken in space. Any major redesign will have to be done back on Earth.

SECTION 18-2 TRIP TO THE MOON

The trip to deep space tests out the spacecraft design. It gives us the opportunity to take off from the Earth and to land right back upon the Earth. We will come down backward riding the beam downward while gravity pulls us down. The front beam will also be used to prevent the spacecraft from tilting. However both beams must be turned on and off so the spacecraft will orbit over a constant point.

The spacecraft can land on any hard rocky surface. It could also melt various surfaces and produce it own landing floor. For the Earth one or more sites will be used and precision corrections must be made during the spacecraft flight. We could bring the spacecraft to a perfect stop thousands of miles from the Earth and watch the Earth rotate until we are ready for a landing.

The proton thruster engines give us complete control. Since they are purely electrical engines with few moving parts, they are extremely reliable and easy to turn on and off. We can take off easy and land easy. All we need is a strong floor to handle our weight. You don't want to sink into ice or land on soft soil.

The time to reach the moon and the peak velocity is found by the simple physics formulas for constant acceleration. They are:

$$S = \tfrac{1}{2} At^2 \quad (18\text{-}1)$$

$$V = At \quad (18\text{-}2)$$

The moon is approximately 240,000 miles away. Half the distance is 120,000. The time to achieve half the distance is found by equation 18-1. This amounts to 6290 seconds when the acceleration is $32 \text{ft}/\text{sec}^2$. It takes 1.75 hours to reach half the distance to the moon. The speed is found from equation 18-2 to be 137,400 miles per hour.

At the halfway point, the front thruster must produce a deceleration of 1G. The rear thruster will be reduced to a small value to maintain stability. It will take another 1.75 hours to reach the moon. The entire trip will take 3.5 hours in comfort. Of course it will take time to turn around and land. At least an additional hour is required during landing.

The only discomfort will occur during the changeover from acceleration to deceleration. The people will become weightless for a few seconds and the seats will be turned around automatically. The people must be strapped into their seats during that time. The entire inside of the ship will become the mirror image as far as the passengers are concerned.

Prior to landing, the ship must turn around. Another period of floating will occur. All these periods should take only a few minutes. The crew of course will be quite used to weightlessness for short periods of time. However some will have to be trained to work under weightless conditions for emergency repairs and the like. The passengers will merely have to remain strapped in their seats during this time. A special room could be set up for those passengers who wish to experience weight-less-ness for awhile.

The trips to the moon will become common place. It will take no longer to get to the moon then it does to go from New York to California. Therefore Hotels will be set up on the moon. Everything must be brought in but the proton thruster engines will enable solid rock to be cut and fused together. We will then be able to build domes and bring water and supplies with us.

There will be constant daily trips to the moon. People will honeymoon on the moon as well. Advanced technology will permit the production of water from rock. The proton thrusters can also turn lead into gold without much difficulty. Advanced machines will convert useable materials into oxygen and water. This will start to happen as soon as the proton thruster engine is developed. Then we can convert useless materials into useful materials and water.

Of course the advanced technology will also turn radioactive waste into harmless products. The technology will move us well into the 21st century. Right now we are still in the 20th century.

The trip to the moon will enable us to learn our spaceship's capability. It will enable us to develop many useful survival techniques such as converting rock into water. The conversion of protons into pure photon energy enables us to do marvelous things, which we could not do before.

We will develop space stations on the moon with sufficient supplies to keep several hundred people alive for up to one hundred years. Cargo ships will bring a lot of water to the moon and we will produce a small city in the rock of the moon with sunlight coming in. We can then grow fresh tomatoes and other vegetables.

SECTION 18-3 TRIPS TO MARS AND VENUS

Mars is approximately 142 million miles from the sun and Venus is approximately 67 million miles from the sun. We are 93 million miles from the sun. Although orbits vary, we can assume that at some time we will be approximately 55 million miles from either of them. Astronomers and the like know the actual distances. However for the discussion here, it is only necessary to understand the approximate distances. The time to reach 27.5 million miles at 1G would be 26.5 hours with a peak speed of 2.08 million miles per hour. The entire trip to Mars or Venus would take 53 hours or 2 days and 5 hours.

Of course, the exact time of the trip depends upon the orbit and the time of year. The trip to the moon could be a day trip. Many people can go. Millions of people can eventually make the trip to the moon each year. It will not be an expensive trip. The trip to Mars and Venus will be much more expensive for the average person. It will be more for the scientist although some people may want it as a honeymoon trip.

The trips to the moon will bring in a lot of revenue since they are short and so many people can go. The trips to Mars and Venus will be interesting as we search for prior life or possible present life-forms. We may look for new elements not found on Earth. We may look for perfect diamonds and other rare materials. Primarily, however the trips to Mars and Venus are for the study of the universe.

We will also build survival space stations on Mars and Venus and maintain a population of several hundred people at all times in case of the destruction of the Earth. We will have sufficient spare spaceships on Mars and Venus and be ready to take off quickly toward outer space in case of a period of intense solar radiation from the sun which could destroy all of mankind close to the Earth, or the inner planets. Such future radiation could last for several years.

SECTION 18-4 TRIP TO PLUTO

A trip to Pluto should be undertaken. Pluto is 3660 million miles from the sun. It will take 12.7 days to reach the halfway point and we will achieve a peak speed of 24 million miles per hour. The entire trip will take 25.4 days to get there. This is a good test trip. It will enable scientific study of the universe from far away from the Earth. Conditions will not be very pleasant on Pluto but the trip will be worthwhile. Hopefully the surface will be solid enough to land

If possible we will build a space station on Pluto with sufficient supplies to maintain life for one hundred years. Pluto should be very safe from any possible radiation from the sun or long periods of destructive meteor showers upon the Earth. Still it may be necessary to venture out to deep space to insure the survival of mankind.

SECTION 18-5 TRIP TO REACH MAXIMUM SPEED

Our brave astronauts will have to test the ability of man to achieve high speeds and the ability of man to reach deep space. We must test the limits. We must find out how far we can go from the Solar System and still survive. We will head toward Alpha Centauri and turn back when necessary.

In order to reach Alpha Centauri we need to achieve 40% to 50% of light speed. At fifty percent of light speed, the trip of 4.3 light years will take 8.6 years while at forty percent of light speed it will take approximately 11 years. It takes 177 days to achieve 50 percent of light speed when we accelerate at 1G. We may experiment during the trip with somewhat higher G forces such as 1.5G.

At the conservative 1G it will take us six months to achieve 50% of light speed and another six months to slow down. If we traveled in a straight line then the trip would take one year to reach zero speed again, one half year to reach half light speed on the way back, and then another six months to reach the Earth. The alternative is to make small corrections all the time and to go on a very large circular path.

We must keep the accelerations on the circular path very low. The result is that some fancy pathway with various speeds will have to be selected. This will reduce the trip from two years to a little over one year depending upon the tradeoff between sideways acceleration and time in space. We may be able to cut the time to a total of 1.5 years for the round trip in this manner.

Let us look at the following chart for the variation of gravitational shrinkage and inertial expansion with respect to increasing light speed.

Velocity	Einsteinian Gravitational length	Doppler forward length	Doppler rearward length
0.0C	1.000	1.000	1.000
0.1C	0.995	1.105	0.904
0.2C	0.980	1.225	0.816
0.5C	0.866	1.732	0.577
0.9C	0.4436	4.359	0.229
0.99C	0.141	14.107	0.0709
0.999C	0.022	44.710	0.0224

The Einsteinian gravitational length is the shrinkage of distance with velocity as per the space-time equations produced by Einstein. Unfortunately his relativity did not account for the Doppler differences. The length of an object decreases in the rearward direction of motion and increases in the forward direction of motion. This produces differences in the forward direction and the rearward direction for an object in motion.

These differences are not important at low velocities such as we have upon the face of this Earth. When we attempt to bring a spaceship upward toward the speed of light, objects get distorted. At fifty percent of the speed of light a 6-foot person becomes over 10.4 feet in the forward direction but only 3.5 feet in the rearward position. A ceiling of a spaceship compartment, which is twenty feet, becomes only 10.4 feet in the rearward dimension. We must make sure the ceilings are at least 20 feet high between compartments on the spaceship.

When we look at the workings of our own body we find that our organs will struggle really hard when we attempt to reach 50% of light speed. Einstein did not understand modern Doppler radar techniques. He used a root mean square approach to his space-time. This works well for orbital motion but fails to work for ordinary classical linear physics motion.

The human body will start to fail at 0.5C or less. The Einsteinian elevator problem was incorrect. Instruments will sense constant velocities as soon as they become large as we approach the speed of light. This is unfortunate since it appears we are destined to live only within our solar system or slightly beyond.

As we look at the chart we find that a spacecraft traveling at 0.1C has a slight Einsteinian shrinkage of one half of one percent. This is small. The spacecraft looks ten percent larger in front of it and ten percent less to the rear of itself. A man would experience the same distortions. When we go up to 20 percent of light speed, the Einsteinian gravitational shrinkage is 2 percent. The forward length is 22 percent more and the rearward length is 18 percent less. These distortions are getting large.

We see that it is very difficult to bring a spacecraft much beyond one-half light speed without severely endangering the life of the crew and the passengers. If we brought the spacecraft up to 0.999C, the six-foot man would look over 270 feet long in the forward direction and only 0.1 feet in the rearward direction. The human body will fail with such great distortions. The spacecraft will fail as well. It is only primary particles, which can endure such high speeds.

Of course future man may find some way of producing a force field so that the distortions will be eliminated. Higher man with super advanced technology may be able to overcome the distortions caused by the forward Doppler length and the rearward Doppler length. Right now all we can do is to test the limits of spacecraft and people aboard and see when they experience discomfort. Hopefully we can achieve between 0.4C and 0.5C. This will make a trip to Alpha Centauri a possibility.

A very conservative estimate would be that we would be lucky only to achieve 0.2C. This is no problem for solar system travel but not good to achieve the stars. We still can achieve Mars or Venus in two days. We can also travel into deep space toward Alpha Centauri for safety during the destruction of the Earth.

SECTION 18-6 PREPARATION FOR THE VOYAGE

As we look into the far future, a time will be reached where the life of mankind upon the Earth will degenerate. The over-population of mankind will cause all the resources of the Earth to become used up. The quality of life decreases with increasing population. Mankind produces more and more hungry poor people as time goes by.

The fish in the sea will become depleted. The rising waters on the Earth will drown many people. Storms will become quite strong. We will see hurricanes of 400 miles an hour and tornadoes of 500 miles an hour. The virus's and bacterial infections will start to kill off hundreds of millions of people at a time.

We may find either dramatic events, or slow changes in the future, which signal the death of man upon this Earth. We must always be prepared for the final times. It may be atomic war. It may be one country attacking another with biological weapons intended to destroy all of mankind except them.

In spite of all this, an International agreement must be reached to insure the survival of the species. Even though we kill each other off for food upon the Earth, mankind must agree not to bring their wars of survival into space.

In order to insure international cooperation, each country must select the best of their peoples to participate in the survival of the species. This will be similar to the Olympics. Each country shall select their best scientists, doctors, educators, plumbers, farmers, housewives, etc. to take part in this great effort to save mankind.

Teams of these people will be sent to the Moon, Mars, and Venus. Some will be maintained in space heading toward Alpha Centauri. The teams will be armed and protected against terrorists who might want to destroy all of mankind. This International force will be composed of the best cross-section of humanity. All candidates will meet the approval of the International Candidates Society.

In general, the United States will take a leading role in the operation. It will protect the project. Times will be bad far ahead from now and although all of mankind will be wiped out, it is necessary to preserve the bodies of man so that higher man can be produced. It is a joint effort of mankind. Everyone has a stake in this project. If it fails then higher man may not be produced and the entire suffering of mankind to date will have amounted to nothing at all.

To make matters worse, all of mankind will enter the final world of the dead with no where to go. It will inflict a very great collective pain upon the souls of the masses of mankind. It is bad enough that the future looks very bleak but mankind needs hope. We need the ability to fulfill ourselves in life. Even though we battle and kill each other, a time comes when all men must strive together as one.

When times are good people will travel to the moon and the planets for vacation. When times are bad, the moon and the planets will become restricted except for the International Candidates. They will embark from a landing pad in the United States or some neutral country. They will be protected by International troops. The United States will take the primary responsibility for the task of the protection.

All the candidates will be trained in English. No one who leaves the Earth will have any problem communicating with anyone else. There will be an equal distribution of men and women. Many Children will be chosen as well. They will go with their parents or guardians. The children are necessary since it may be as long as 40 years before they return to the Earth. They must be able to produce higher man as their offspring. Therefore the very young must go along as well.

Of course modern medicine will permit women as old as 60 to carry children as long as their eggs are stored. The spaceships may also carry the eggs and sperms of selected candidates as well. The system is independent of the exact people involved. They will merely be the carriers of higher man. Their children will be born from the highest spiritual purity of the Human Life-force. They only need to provide the body to carry the children and the eggs and sperms for the basis of the evolutionary process.

The candidates will serve as the new Chimp/Apes for the New Earth. They will become the carriers and producers of higher man. They will return to the New Earth and build a city. The candidates will be people from all over the Earth.

Yet, they will have a single purpose. They will actually see the evolutionary process working.

All candidates must be of good moral quality and dedicated to the collective team effort of survival. It will be an International Olympic team for the survival of mankind. We want the highest quality of moral spirit and dedication from the candidates.

Most of the candidates will be teenagers and young adults. Some will be married. Others will be single. Some will be children. They will all be part of the greatest adventure of all times. Some will fall in love. Others will fall out of love. In spite of this they must accept their common purpose. They must be dedicated to the group first and themselves last. We need the romantic idealism of youth. However older people who will not survive the entire journey will serve as well. They will be the teachers and educators. They will be the doctors and the ship's commanders. Again people of utmost dedication will be required.

We do not know when the final time will come for man. Yet when it starts there will be many signs. We will see great deaths all around us. At that time, we must insure that the candidates are off the Earth. Some will be kept on the Moon and the planets. Others will be kept in deep space. The Hotels on the Moon and the Planets will be reserved only for the candidates and whatever military escorts are necessary.

It will be necessary to destroy any space ships attempting to take over the International complex on the planets and the moon. The military forces will have to attack and destroy anyone who attempts to force themselves upon the selected of mankind. There are those despots who might attempt such things. They must be destroyed. The military forces must also be prepared to destroy any incoming missiles heading for the hotels from despotic countries bent upon nihilism. We then may have a small war of the worlds as the International forces battle despotic countries bent upon the total destruction of all mankind. We have come to understand that some people exist whom only want to kill and destroy others. They must be stopped.

The proton thruster rockets will get us to the moon readily. Unfortunately they also could serve to destroy our Hotels on the Moon from the Earth. We always must be prepared to head out into deep space as the only hope for the survival of mankind. Once in deep space, our spaceships will not be vulnerable to attack except by other spaceships from the despots of the world. Yet the International force upon the Earth should be able to destroy any enemy spacecraft heading outward to destroy the International candidates. In addition, the rear proton thrusters can destroy any missiles coming toward the spacecraft from behind.

CHAPTER 19: THE VOYAGE TO THE NEW EARTH

SECTION 19-0: INTRODUCTION

Someday in the near or far future, the Earth will experience a period of destructive forces, which will make life very difficult. These forces may be man made or natural events. Man must be ready to survive meteor strikes, atomic war, huge plagues, intense solar radiation, and the self-destruction of the Earth by an Einsteinian oscillation.

The Earth can go off balance, and break into an Einsteinian oscillation in which orbital speed changes with a resulting change in the period of the year and the day. The oscillation was part of the original production of the Earth's orbit. An unbalanced condition on the Earth could set off another oscillation.

This would cause ocean waters to rise upward with very destructive force. The trade off between orbital energy and rotational energy will destroy the Earth as we presently know and rebuilt it into a New Earth. A huge space rock could precipitate the oscillation. Likewise a strong disturbance from the sun could cause the oscillation as well. The Earth can also be destroyed by a period of intense meteor strikes. We must be prepared to survive in space while the Earth is operating in a very destructive pattern. Finally man himself could precipitate a huge atomic war which destroys all life for a long period of time.

The huge destruction of all of mankind produces tremendous spiritual pressure and suffering in the world of the dead. This will cause evolutionary changes to the offspring of the few survivors. The death of evolved man results in the production of higher man.

We do not know when the events in this future story will take place. Man has been upon this Earth for over a million years. We could last one thousand years, ten thousand years, or perhaps a million more years. However someday we will self-destruct. In this chapter we will look at a little story of future times.

In this chapter we will look at the trip toward Alpha Centauri as a final atomic war is destroying all life upon the Earth. We will then look at the return trip to the New Earth. In addition, the massive destruction caused the Earth to break into a space-time oscillation with resulting changes to the orbital patterns. We will look at excerpts from the diary of a historian on one of the spaceships circling the moon.

SECTION 19-1 THE WAR ON THE EARTH

"Some time has passed, there have been several generations of candidates for the survival of mankind living on the Moon, Mars, and Venus. We have found out that we can reach 0.45C before getting ill. We also found out that we can achieve half way to Alpha Centauri and back without problems. We have not tried to go all the way yet. At the present time the situation on the Earth is very bad."

"The Nations have been battling each other for food and farmland. There is little fish in the sea and no rainforests exist at all. There has been an atomic war going on for quite awhile. The Earth looks quite horrible. People have been using biological weapons and entire countries no longer have any living people in them. Everywhere we find death and rot."

"Everyday in the telescope we see that the Earth trembles more and more. The volcanoes erupt. The seas rise. A billion people on low-lying areas have already perished. The sky is starting to darken. There is little sunlight in many parts of the world. It looks like man will be no more. Things are not looking good at the moon station either. One nihilist despot sent several missiles with atomic bombs to the moon. We lost one space station and half of our candidates. "

"The rest of the candidates are now circling the moon in the space ships. It looks like we may be attacked again. It is necessary to move outward toward Alpha Centauri until the war on the Earth is no more. The people no longer have long to live. It is only a matter of months before no one is left alive. We had expected the worse yet, when it came, it was unbelievable."

SECTION 19-2: THE TRIP TOWARD ALPHA CENTAURI

"We are all heading out to deep space. There is no use circling the moon much longer. The enemy has tried to destroy us with an atomic laser beam. One of our ships was damaged but the crew was okay. It looks like they can harm us. Therefore we are heading out."

"The ships departed the moon base. The same is true of Mars and Venus. While the enemy is still alive upon the Earth, the ship's commanders decided to leave the area. One set of ships already was one light year away. Now all ships decided to leave."

"All the ships are prepared to survive at least ten years in deep space near Alpha Centauri. They are all heading out to insure that no despot will send missiles toward them. The voyage to Alpha Centauri began for us. Each day we look at the Earth. Each day the dust cloud cover gets worse. The radioactivity readings from the Earth are not pleasant. There is a cloud of radioactive dust all over the place. This is killing everything. It looks like one gigantic death wish."

"The Moon was attacked several more times. It does not look too good there. Yet without an atmosphere, things should be okay to return to the moon within a few short years. Our spaceships are prepared for many years in space. Right now it looks like the space stations on Mars and Venus stand intact. We will be able to return to them as soon as things are safe.

"Some time passed and a missile was spotted coming right toward us. It did not look good. Fortunately it came right up our tail. We turned our back engine on to a sharp beam and managed to destroy it before it got within a hundred miles. Now we have to look very carefully to see if something else comes our way."

"Some time passed again and the Earth is completely covered in a dust cloud. It is one horrible radioactive dust cloud. It is shaking violently as well. The entire Earth is undergoing a transformation. The moon is being affected but not severely. Things look okay on Mars and Venus. We are now 2 years away from the Earth. The images we get are about one year old."

"Some people want to head back toward the moon. The danger appears to be over. The Captain feels that we should continue outward slowly for another year. The universe is quite beautiful from deep space. Everyone is very sad what happened back home. We do not know if we will survive to come home."

"It has now been three years in space. The Captain has decided to turn around. He will slow the ship even more and turn slightly sideways until we are pointed toward Mars."

"We are all weary of being on the ship. A few people have died but most have survived. It has now been six years and Mars is right up ahead. The Earth looks quite horrible. The radiation readings are getting less and less. It is just covered with a horrible black cloud."

"Our ship is landing on Mars. A fellow ship is landing on Venus. Things are okay. The people feel good to be landing upon a planet again. The enemy was not successful in damaging the space station and hotels on either planet. It will take a little work getting it back into working order but everything seems okay.

The stored water and food on Mars is in excellent condition. We started the atomic generator up and the little city lit up. It was a great feeling to be on land. The constant acceleration (1G) on the trip caused most people to feel okay. Now they have to get used to Mars. Yet it was found in the past to be okay."

"It has been over a year on Mars. We study the Earth in the telescope each day. It looks so bad but the atmospheric radiation levels are down. We are hoping that these levels will decay so that some place on the Earth will be safe to land."

"We have been on Mars for another five years. The radiation level on the Earth appears acceptable. The clouds are horrible but here and there a small break occurs. When the sunlight is just at the right angle we get a glimpse at the ocean below. It is muddy and horrible but at least it is water."

"Another year has past and the Captain wants to travel to the Moon and restore the Moon station. Life on Mars has been okay. We are used to eating the horrible bland food. Now we can take off. We still have three ships with us. We will leave some people on Mars and the rest will travel to the Moon."

"It took only a little over two days to get to the moon. The radioactive levels near the space stations are okay. Things are a mess there. Things are better at our storage facilities. There are many machines in good condition. There is also a hundred years supply of special canned foods. Now we can eat fruits and vegetables, which were radiated and then frozen, on the far side of the moon. Everything appears great."

"The Captain wants us to repair and rebuild a space station. This is going to take us a year. We have plenty of supplies and the spaceship itself will do a lot of work for us. Luckily we have stored compete spare parts to rebuild a small station."

"The work has been completed. It took over a year but now it is done. We have been joined by another spaceship. We now have 500 people working on the project. There are still another thousand people who landed on Mars and Venus after us. They will stay there for awhile."

"The time has come to land upon the Earth. It is only a few hours trip but it is the most frightening thing. It is going to be dark. The clouds are still so black."

"We are coming down now. Our radar says that we are heading toward the top of a mountain. We cannot see a thing. Fortunately the proton thruster rocket brings us straight down. Our instruments tell us that land is up ahead. Nothing is visible. We will land. We were lucky. We hit solid rock."

"The radiation measurements are not bad where we are. The Earth is full of poison gas. There is oxygen but there is so much poison gas. Yet we have landed. No one is alive on this Earth except us. The water is full of poison as well. Yet we can purify it readily."

"We now have fresh pure beautiful water. Our scientists are studying the condition of the air. The computer program says that in several more years, the air will self-cleanse. We produce our own clean air aboard ship but now we can pump and then cleanse outside horrible air. We are now breathing purified Earth air. It is a great thing to breathe the air of the Earth again. The purifier worked well and the air has a beautiful saltwater smell as compared to our bland normal air. We are excited by the smell of the Earth's air."

"We are filling up with water. We are taking some soil samples as well. We will return to the Moon. We will wait a few more years before returning."

"It has been another five years on the moon. We have not had any success having any children on the moon. Some people have died but the majority of people have survived. We move from the moon to Mars where the gravity is 39% of Earth's gravity and Venus where the gravity is 88% of Earth's gravity. A lot of people like the planets better. They are more comfortable. On the moon we feel so light but we are able to enjoy it. We love the clear sky on the moon. The view of the universe is magnificent."

"Many people have preferred to stay in space. The spaceships have been very reliable and comfortable. The people travel around the solar system. Some people like Mars and Venus. It makes them feel a little like home. Yet they are forced to remain in the space stations. We all do a lot of different things to keep busy. We travel to all the planets and have learned a lot. It has been a great adventure."

"Many of the people are getting older. We must return to the Earth soon. Things are looking better. We see some blue sky here and there. We see some rain clouds. The poison gas is decreasing. The Captain feels we could find a spot to land."

SECTION 19-3: RETURNING TO THE NEW EARTH

"The time has come to return to the Earth. We are all excited. We see a safe spot to land. Our instruments say that this spot is very safe. We are now leaving the moon, hopefully for the last time."

"The Earth is coming near. We see an Island. It is small. We see that some vegetation has returned. It looks like one little Island in the whole wide world exists just for us. The winds are moving just right. This one little spot in the world is beautiful. It is a little oasis in a world of dark clouds."

"We have landed. The air tests good. For the first time in almost fifteen years man has landed on a planet where he can breathe the air. The entire Earth still looks like a mess but this one little Island stands alone."

"The Island has birds and the water has fish. There are trees growing. There are some strange fruits that we have never seen. They taste great. We are testing everything out carefully. Everyone seems okay. The other spaceships are anxious to land. Our Captain has told them to wait another month. They told us they see another Island in the telescope. Our Captain told them to go for it."

"The other space ship landed. The other Island was a thousand miles away. It was good too. The rest of the spaceships want to land as well. Half will come to our Island and half to the other Island."

"The other space ship brought us small speedboats. We have a small electric propeller plane as well. We are taking small trips here and there. We cannot find the normal continents. Everything has changed so much. Things are still changing. Much of the world still lies in poison gas. Yet more and more clear skies are opening up."

"The women are getting pregnant. This has not happened all these years. Some of the little children are now getting married. They are now young adults. The women who were twenty when we left now want to have children as well. There are only about 1500 people on the Earth and there are a lot of pregnant women. Some are carrying twins."

"We have been on the Island for nine months and the first children are starting to be born. The Newborn looks like us but they seem to have a lot of spiritual energy about them. They all seem so beautiful. We don't find any birth defects among any of them. They are all so healthy. The doctors have been studying them. They are slightly different than us. They appear to have better body chemistry. One was accidentally cut upon birth by the surgeon's knife. He healed in very quick time right before our eyes. The children appear to have better healing chemistry."

"As we watch the Children grow we notice that they are really well behaved. They all get along with each other very well. The Children are very spiritual and pure. We eat fish but they do not. They love the beautiful fruit trees, which grow in abundance on the Island. The Island is slowly rising. It is becoming a bigger Island. There is a fresh water stream. More trees and fruits seem to grow each day."

"The rest of the world still looks like a mess but the two Islands provide us with life. We brought corn seeds with us and are planting them. Everything seems to grow fast. We still have plenty of food on the spaceships but as the Islands rise, the food supply gets larger and larger."

"The scientists have been studying the Earth. They feel it may take another hundred years for the entire Earth to be okay. Everything seems to be gone. Everything, which existed before, seems to be gone. We only find poison gas in the atmosphere and fresh soil arising from the seas. The areas of clear sky are increasing and the poison gas is reducing. Birds are appearing. The fish are coming back. Yet it is mostly where we are."

"Life is simple. We all work together. The children are beautiful and the whole team of people who left the Earth years ago worked out. It is an interesting life. The weather is great on the Islands. We are baking clay and mud to make adobe houses. We are building a little city around the spaceships."

"The Earth trembles quite a lot. There are earthquakes all over the world. We feel the Earth shake but our land keeps rising upward. It gets bigger each day. We picked a good spot to land. The land is very fertile and the harvests are very good."

"We are observing an evolutionary process, which is transforming the Earth before our eyes. We see the process on this Island and the next, which seems to have been prepared for us. Somehow the Earth was made ready for us. Everywhere else, it is still pretty bad. Yet the sun shines here often. The rains come and the water is pure from them. We are just in a perfect spot. The other Island is doing as well."

"The mountains on the Earth appear to have been destroyed. The severe oscillation of the Earth destroyed everything. Everything is more flat. Most of the Earth is covered with a thin layer of water. The Earth became more perfectly round. Ice seems to be forming at the poles. Once in awhile a bad storm comes with its poison gas. We run into the spaceships until the danger passes. Yet the storms are getting less frequent. The earth is quieting down. The earthquakes are getting less and less often."

"Everything is destroyed or under water. The sky is getting clearer all over the Earth. We fly over and see the remains of broken cities underwater. Here and there we see a piece of a pyramid or a piece of the Great Wall of China protruding from the water. We see new species of fish emerging in the water. They seem to thrive on the poisonous mess. They appear to eat the horror. They are eating the polluted mess which mankind left behind."

"The scientists are finding new bacteria, which are also growing and eating the horror of prior human existence. Nothing seems to hurt the children. We get sick when we get too close to the horror far from our islands. The children do not get sick. They seem to have better immune systems than we do. They process the heavy metals and mercury without absorbing it in their bodies."

"Some of the people are growing old now. The new children are getting married and starting to have children of their own. Their children are even more beautiful than they are. We are witnessing an evolutionary process leading toward higher man. Each generation is moving upward. There has not been one birth defect in any of the children."

"The children of the children are not like us. They seem to have been born with great compassion toward each other. They seem to have humanity about them, which is unmatched. All the children are very spiritual."

"We do not feel the way the children do. Yet we witness the way they are. Their body chemistry must have a better spiritual connection between their space-time soul and their inner mind. The Children have been born with a great spiritual depth, which we do not possess."

"We are slowly building this Island City. It is difficult. We have been reduced to simple living. We have modern medicine in the space ships. We have some electrical power in the city. However for the most part, the houses go dark at night. In the center of the city we have lights and a library. At the outskirts we have simple homes with no lights."

"Many people are happy with the simple life. The fruits and vegetables are abundant. The new children are vegetarians. They do not like when we kill a fish. Most of us have become vegetarians as well. We do not like to upset the children and now we have no need to eat any meat. The Earth has provided abundantly for us. Gone are the days when all the species of this earth hated and feared man. We can go into the water with the new sharks and other horrible looking creatures. They do not fear us nor hurt us."

"Some of the mammals in the sea come to us for help when they are hurt. They seem to know that we will help them. The children are very good at helping them. They love the children. We now live upon an Earth where nothing seems willing to attack or hurt us."

"Some jungles are growing. The scientists have spotted some poisonous snakes. The children have picked them up. We are afraid. The children are not afraid of them. They are not afraid of the children. It is strange. We no longer have to live in fear of anything. The children say that someday all the poisonous snakes will be gone. They will slowly become more loving creatures. Yet they cannot hurt the children. They can only hurt us."

"We no longer have to live in fear of ourselves as well. There are no locks on any doors. Everybody treats everyone else like his or her brother. The Children get married and they do not cheat on their spouse. Some of us still do. It appears to be part of our nature but not the children's nature."

"There are neither rich nor any poor. We all work together for the common good. We do not have any money. We really have no need for money. We eat together in a community kitchen. We work together and take care of the children together. The children all seem very smart. None of them have any learning disabilities. They are smarter than we are. We are teaching them everything we know."

"Our Captain has decided that I should not teach the children all the horrors of our prior existence. We have destroyed a lot of the history of mankind. The Children know that the Earth was destroyed. They know that we are an evolved more primitive image of man. They know that they are the start of higher man."

"The children are taught math and science and medicine. I teach them about the history of mankind. They seem to understand all these things very well."

"Time has passed; the land has risen higher. The ice has come back on the poles. The days are a little longer. The year is a little longer. We live in a beautiful climate. There are several other islands forming. We now have some mountains a mile high. We are getting new kinds of animals. We see mammals changing before our very eyes. Some have become pets like dogs."

"We are starting to die out. The population of mankind has risen to almost five thousand people. We are finding iron ore and coal and making a limited amount of steel. We are determined to live modestly and make sure that we do not pollute the new Earth. Everything is so pure and beautiful."

"The diary is coming to an end. There are only three candidates left. We are the remaining people who were born on the previous Earth. It has been an amazing voyage. It started in such great sadness. Billions of people died in horror. They fought and killed each other. The earth responded against them as well. Ancient tribal man is gone. We are all one people now."

"So many died. Billions died. The net result was the beautiful children. The beautiful children were born from the purified collective soul of all the dead. The suffering of evolved man has given birth to new and higher man."

"The three of us remain. We leave to the children this Earth. We leave our love and our hope that they will never become like us. They will grow and multiply. They will become a nation of millions. Right now they are merely modern tribal men and women. They are a small tribe of five thousand. Later they will become a civilization."

AT PARTING

This book has taken almost 20 years to reach this point. The manuscript involved thousands of pages starting on an 1898 Underwood typewriter and finally ending up on a modern computer. It evolved two intense years from 1981-1983 where the constants of the universe were studied with a hand calculator and the basics of the Dot theory formed.

The original basic Dot theory remained intact all these years since 1981. The Doppler understanding of the dot is only a 1999 addition. The conversion from mass to charge kept changing as new conversions replaced older ones. About four different conversions were seriously considered all these years. Looking back, it appears to the Author that the first choices were inadequate. Yet, they enabled the work to progress. When you climb a mountain the first time, you take many wrong paths. There are many dead ends along the route. Yet much is learned in the process since the choices tend to be sister solutions of each other.

The present conversion from mass to charge started in 1998 and yields an understanding of gravity and the equation for the gravitational constant. The other conversions did not yield such equations. The inclusion of Doppler space-time has also added fruit to the labor. As we move toward the speed of light, the Doppler length to the right approach infinity and the Doppler length to the left approach the size of a dot. This enables us to see that the dot itself is a Doppler dot. It is the result of a charge Q at the radius of the universe moving at the speed of light away from us.

Long ago, the Author chose to look at the dot from an inverted image perspective. Now after almost 20 years, such a perspective becomes clear. In the past, it was a means similar to a Thevenin equivalent of the Universe. Now it appears as a Doppler equivalent of the universe.

The original manuscripts looked at the universe from many different perspectives. Multiple solutions were chosen and presented. This approach helped the Author to look at all the possibilities. However, in order to keep the book a reasonable size multiple solutions were omitted from the presentation.

The discovery of the Doppler Earth enabled the Author to present an understanding of the driving force for the creation of life upon this Earth. The Doppler Life-force can be looked upon from a religious or purely scientific perspective. It will give philosophers, and theologians, food for thought.

Since the Author has succeeded in bringing the equations of gravity to light, it is time to send out this version to mankind. We all have limited years and this initial study is completed. It has been a great hobby.

APPENDIX -1
DEFINITION OF SYMBOLS USED

SYMBOL	DEFINITION
AC	Alternating Current
A_U	Acceleration of Universe
C	Speed of Light in vacuum
$\underline{C}$	Capacitance
C_D	Dot Capacitance
d	Differential operative
d(t)	Deriviate with respect to time
DC	Direct current
e	2.71828
E	Energy
E_D	Energy of a single dot
E_{P*}	Energy of p* subparticle
E_U	Energy of Universe
ε_o	Electrical permitivity const.
f	Frequency
F	Force
F_D	Force acting upon a dot
G	Gravitational Constant
h	Plank's Constant
I	Electrical Current
I_D	Dot Current
K	Coulombs constant
L	Inductance
L_D	Dot Inductance
M	Mass
M_D	Rest Mass of dot
M_E	Mass of electron
M_{Eo}	Rest Mass of electron
M_N	Mass of neutron
M_{P*}	Mass of P* subparticle
M_P	Mass of proton
M_U	Mass of Universe
N_D	Number of dots
N_N	Number of dots in neutron
N_U	Number of dots in universe
Q	Electrical Charge
Q_D	DC Charge of one dot
R, r	radius or distance
R_P	Radius of proton
R_{P*}	Radius of P* subparticle
R_N	Radius of neutron
R_U	Radius of Universe
t	time
T_U	Time of Universe
U_O	Electrical permeability
V	Voltage
V_D	Dot voltage
X	Algebraic variable
X^2	Square of X
$X^{1/2}$	Squareroot of X
λ	Wavelength symbol (lambda)
1E-27	The minus 27th power of ten

APPENDIX 2
TABLE OF CALCULATED VALUES

QUANTITY	SYMBOL	VALUE
Dot Current	I_D	2.56699E-38
Dot DC Charge	Q_D	1.42186E-60 cou
Dot Energy	E_D	1.2329E-55
Dot frequency	f_D	2.5497E-20 / sec
Dot Inductance	L_D	2.3516E21
Dot Impedance	X_O	376.828 ohms
Dot Period	T_D	3.9220E19 sec
Dot Period	T_D	1,243.78 Billion yr.
Dot Voltage	V_D	9.67065E-36
Radius of universe	R_U	1.48901E26 meters
Radius Subparticle	R_{P*}	1.45712E-15 meters
Radius Neutron	R_N	1.31959E-15 meters
Radius Proton	R_P	1.32142E-15 meters
Mass of Dot	M_D	1.37188E-72 kg
Mass of Hydrogen	M_H	1.67353E-27
Mass of Universe	M_U	2.00559E53 kg
Proton Voltage	V_P	1,089,700
Subparticle energy	E_{P*}	1.6891E-53
Time of Universe	T_U	4.96682E17 sec.
Velocity Bohr	V_B*	1.05367E-28
Wavelength N	λ_N	1.31959E-15
#Dots/charge	N_Q	1.12682E41
#Dots/Subparticle	A*	137
#Dots electron	N_E	6.64008E41
#Dots Neutron	N_N	1.22090E45
#Dots protons	N_P	1.21922E45
#Dots/ subparticle	A*	137
#Dots Universe	N_U	1.46193E125
# P* universe	N_{P*}	1.06710E123
#Neutrons Universe	#N	1.19742E80

APPENDIX 3
STANDARD VALUES

Quantity	Value (5 places)
R_{BOHR}	5.29177E-11
C	2.99792E8
(Fine Const)$^{-1}$	137.036
G	6.67260E-11
ε_o	8.85418E-12
h	6.62608E-34
K	8.98756 E9
Mile	1.609km
M_N	1.67493E-27 Kg
M_N	939.565 MEV
M_P	1.67262E-27 Kg
M_P	938.272 MEV
M_E	0.910939E-30
M_E	0.510999 MEV
Q	1.60218E-19
U_O	1.25664E-6
Z_O	376.731

INDEX OF SELECTED TOPICS

Acceleration of Earth	154, 173,174	Electron Energy	62
Acceleration of Universe	2	Electron Frequency	62
Binding energy of proton	66-70	Electron's Inductance	62
Bohr acceleration	154	Electron's Capacitance	62
Bohr Angle A*	51	Electron's Impedance	62
Bohr AC & DC currents	64	Electron's AC current	62
Bohr AC & DC energies	64	Electron's DC current	64
Bohr Delta Energy levels	61	Electron's wavelength	61
Bohr Energy	60,61	Electron's AC voltage	62
Bohr Inductance	63,64	Electron Magnetic Repulsive Force	65,66
Bohr Electron Repulsion Current	65	Electron Radius in Bohr Orbit	49
Bohr Orbit Complete Equations	148,149	Electron Radius in Neutron Orbit	47,48
Bohr Orbit Equations	65	Electron velocity in Bohr Orbit	49
Bohr Orbit space-time equations	124-126	Electron Velocity in Neutron Orbit	47,48
Bohr Orbit energy	60,61	Electron mass in Bohr Orbit	49,50
Bohr radius	27,45,50,61	Electron Mass in Neutron Orbit	47,48
Bohr Expansion Velocity V_B*	27	Energy of Universe	3
Charge Formula	128-130	Energy of Red photon	56
Charge invariance	89	Energy vs. Velocity	92,133
Chart- Quantum Numbers	123	Fast Forward Law	120
Cosine A*	51,52	FM Doppler freq.	134
Cubic length	110	FM Doppler universe	138-141
Cubic light speed	110	Force of Universe	3,6,34
Cubic time	110	Frequency Invariance	111
Cycle time of Universe	30	Frequency modulation	134,135
Delta Clock time	113,114	Frequency of Red photon	56
Delta Light Speed	39	Frequency of proton	53
Delta Radius of Universe	40	Figure- Grav. Force right hand rule	163
Delta Time of Universe	40	Figure-Gyroscope rotation	109
Doppler frequency left, right	73,111,114	Fine Constant	24,41,51
Doppler Gravity	177,178	Gravitational constant	6,22-28
Doppler Gravitational mass	73-75	Gravitational Constant Formula	151,152
Doppler inertial equations	73-75	Gravitational energy	92
Doppler inertial mass	74,75,83	Gravitational Force	26,152,153
Doppler inertial length	88	Gravitational Rules	170-177
Doppler Length $-L_{XR}, L_{XL}$	88,100-102	Gyroscope	108-110
Doppler Mass CG	84	Image Time	94
Doppler mass – front, rear	83,85,97	Impedance of Universe	9
Doppler RMS Freq.	111	Inertial laws	98,99
Doppler space-time principles	71-75	Law of Charge production	70
Dot Capacitance	30,32,33	Law of Images	169
Dot Current	9,10,32,33	Law of Inertia	98,99
Dot Charge	15,16	Law / Rule of Motion	146
Dot energy	10,18,28	Law of Orbital Magnetism	150
Dot Force	10,34,35	Law of Proton/Proton Attraction	66
Dot Inductance	30,33	Law of Proton Red shift	146,147
Dot mass	10,11,12,13,16,18,41	Law of Spin Magnetism	150
Dot power	10	Law of Universal gravitation	169
Dot voltage	9,14,33	Law of Virtual Charge	68
Dot Wavelength	43	Law of Red Shift	60
Double slit	118-121	Length Invariance	88
Dual Doppler Earth	179-181	Length Formulas	127
Electron de Broglie Wavelength	49,61	Length X, Y, &Z (GG/MM)	95,96

Lightspeed Equations of Universe	24,25	Photon's radius	57
Light speed hysteresis loop	38-40	Photon's relative speed	122
Lightspeed Variation over cycle	39	Proton Wavelength	53
Mass formula	130	Proton Radius	44,46,53,54
Mass conversions	19,21	Proton's AC current	55
Mass increase (classical)	102	Proton's AC voltage	55
Mass of Hydrogen Atom	151	Proton's binding energy	66-70
Mass of photon	118,119	Proton's capacitance	53
Mass of Proton	44,48,55	Proton's inductance	53
Mass of Neutron	49,55	Proton's voltage	14
Mass- X, Y(GG/MM)	73-75	Radius of Universe	2,3,4,12,43
Mass of Universe	3,4,11,12,13	Ratio L_X / L_Y (GG/MM)	96
Mass of Universe at big bang	136,137	Radius of p*	41,42,46
Mass (Differential)	50	Red photon's AC Current&Voltage	57
Mass Transient Equation.	94	Red photon's capacitance	57
Minimum Radius of Universe	36	Red photon's energy	56,60
Multi-light-speed chart	131	Red photon's inductance	57
Neutron expansion velocity V_{N*}	28	Rotational Doppler rules	104 –108
Neutron Orbit Equations	47,48	Space-time gravity	157-162
Neutron Radius	46,53	Space-time math	31
Number of charges per electron	15	Space-time quantum mechanics	123,124
Number of Charges per proton	15	Three dimensional forces	145
Number of dots in electron	15	Time invariance	89
Number of dots in Neutron	13,15,42	Time X, Y, & Z (GG/MM)	78,112,113
Number of dots in Proton	15	Time of stability	94
Number of dots in Universe	13,41	Time of Universe	3,4,12,27
Number of Neutrons in Universe	13	Velocity (Differential)	78
Number of p* in Universe	41	Velocity of Light in glass	119
Number of p* per neutron	42	Velocity of Neutron Expansion V_{N*}	28
Orbital Clock distance formulas	112,113	Volume of Universe	41
Orbital Clock frequencies	111-114	Wavelength of electron	49,61
Orbital clock- Time formulas	112,113	Wavelength of Neutron	49
Plank's Constant	23,24	Wavelength of Photon	48
Phase Time	141-144	Wavelength of Proton	44,48,53
Photon Mass conversion	119	Wavelength of Red Photon	56

NOTES

ORDERING INFORMATION

This book may be ordered at any time directly from the publisher. A supply will be maintained so that it will always be in print.

The book price is $25 plus $3 shipping and handling for surface postage. For books sent to Virginia addresses, please add 4.5% sales tax.

Surface postage is by truck for U.S. and Canada. Surface postage is by ship for anywhere else in the world. Allow 8 weeks for ship delivery to foreign ports.

Add $3 for U.S. and Canada priority or airmail.

Add $9 for global priority or airmail.

All payments must be by check, money order, or international money order payable in U.S. Dollars.

Send check or money order to:

>Starway Scientific Press
>
>Post Office Box 62174
>
>Virginia Beach, Virginia 23466
>
>Email: starway@infi.net

For additional information write to Starway Scientific Press at the above address.